Voice Teletraffic
Systems Engineering

The Artech House Telecommunication Library

Voice Teletraffic Systems Engineering

James R. Boucher
Artech House

Library of Congress Cataloging-in-Publication Data

Boucher, James R., 1932—
  Voice teletraffic systems engineering.

  Includes index.
  1. Telecommunication — Traffic. I. Title.
TK5102.5.B684   1988   621.38   88-24226
ISBN   0-89006-335-4

Copyright   ©   1988
ARTECH HOUSE, INC.
685 Canton Street
Norwood, MA 02062

International Standard Book Number: 0-89006-335-4
Library of Congress Catalog Card Number: 88-24226

10   9   8   7   6   5   4   3   2   1

To Barbara

Contents

Preface

This book was designed for the use of students, technicians, engineers, and others involved in the specification, design, or administration of modern telecommunication networks. The complexity of these networks dictates a total systems engineering approach to teletraffic engineering. Each component of the network, whether hardware or software, has its effect on the ability of the overall system to perform its task of carrying traffic. Therefore, this book discusses network components, boundaries, inputs, and outputs; explores the internal structures of the various hardware and software components; uses mathematical models to determine equipment requirements; considers alternative system components and methods; provides criteria for specifying the optimum implementation (arrangement); and draws upon experience derived from existing systems, coupled with statistical data and trend analyses, to assist in planning the systems of the future.

Most of the material in this book was derived from extensive notes prepared for use in conjunction with a Northeastern University continuing-education teletraffic engineering course. The comments of students who took that course have led to periodic updates of those notes, and revealed the need for a single document that can be used as a basic textbook as well as a reference handbook. Therefore, it contains introductory material for individuals new to telecommunication systems; mathematical formulas and equations, plus the assumptions upon which they are based; tables for the most commonly used formulas; computer programs to assist in the solution of traffic problems; and typical examples to demonstrate the use of the formulas, equations, programs, and tables. References are provided for those seeking additional background information or more rigorous derivations of the traffic formulas.

I am deeply indebted to the staff who developed and conducted the GTE Traffic Engineering School held at Norwalk, CT, in 1974. The notes for that course were my introduction to teletraffic engineering and inspiration for the course I teach. Excerpts from Paul S. Harrington's presentation on common-control switching systems plus Floyd L. Vanderwater's presentation of planning and forecasting techniques appear herein with their gracious permission.

In addition, the review and critique of the manuscript by my colleagues G. William Carter, Edmund A. Harrington, Louis K. Pollen, and James L. Rollins are sincerely appreciated.

J. R. BOUCHER
APRIL 1988

Chapter 1

Introduction to Telecommunication Systems

This chapter presents an introduction to and overview of telecommunication systems and networks, from the traffic engineer's perspective, using the North American public switched telephone network (PSTN) as a model. It is not within the scope of this book to provide a detailed analysis of the PSTN—the book will simply present the network's role in and effect on traffic carried by its components. For more detailed information on telecommunication theory and systems, *Engineering and Operations in the Bell System* (Bell 1983), *Digital Telephony* (Bellamy 1982), and *Future Developments in Telecommunications* (Martin 1977) are recommended reading.

1.1 SCOPE OF TELETRAFFIC ENGINEERING

The main components of telecommunication networks are generally classified as switching and transmission equipment. *Switching equipment* connects inlet and outlet terminations to each other and to service circuits. They have taken various forms from simple manual cordboards to modern electronic stored-program common-control systems. *Transmission equipment* interconnects the switches with each other and to their terminal equipment. The trunk groups that provide the interswitch transmission paths may be as simple as wire pairs or as complex as time-multiplexed radio or optical links. However diverse, all of these system components have one thing in common—they were designed to handle traffic in the PSTN with an acceptable grade of service.

In processing traffic through the network, we must ensure that an accurate number of appropriate facilities are in the right place at the right time to be effective. The fundamental traffic engineering functions are determination of parameters, traffic system design, and preparation of the traffic order (TO).

1.1.1 Determination of Parameters

Determining the traffic parameters and constraints to be applied to the system under design is a fundamental activity of paramount importance. Generally speaking, these design parameters may be categorized as traffic characteristics and service objectives.

Traffic characteristics include such items as holding times, calling rates, subscriber characteristics, usage, and type of distribution. Service objectives define the degree of congestion and the criteria by which it is to be measured. It is important to recognize that these parameters, as well as the design process itself, are time dependent. Therefore, forecasting these data becomes a critical aspect of the traffic system design process. We are endeavoring to design tomorrow's systems today, using yesterday's data.

1.1.2 Traffic System Design

Once parameters are defined, we may proceed to design the most cost-effective system arrangement. This must include network considerations and properly recognize the effect of current arrangements as well as future plans. Traffic system design refers to a process, just as the determination of parameters is a process, or work function. It involves several disciplines, such as electronics, computers, mathematics, statistics, probability, queuing theory, reliability, and economics.

The traffic system design process concerns itself with the entire system. It must address the fundamental properties of system operation; identify subsystem objectives; and resolve any conflicts between various parts of the system. In a trunking network, for example, it must be concerned with the capabilities and capacities of each switching exchange. Within a switching system it must ensure that the provision of equipment groups is compatible. Common-control switching systems introduce several areas that require a total systems engineering approach to properly coordinate the overall operation of the system.

Traffic system design searches for the optimum solution, that is, the arrangement of equipment and the quantity of devices that minimize the cost of the system while meeting constraints imposed by the system parameters. In the language of operations research, this is the allocation of limited resources. This implies that we are not content with the *status quo* or mere improvement. Thus, traffic system design is involved with the configuration and dimensioning of switching and transmission equipment; the timing of additions; and their effect on service and cost.

The function of traffic system design is not limited to the switching equipment. An equally important area is that of trunk network engineering. The process must consider homing arrangements and network routing rules. The objective is to develop routing strategies that can handle the requirements and meet the objectives with least cost. That is, to configure the building blocks of equipment in the most economical manner, ensuring that service objectives will be met, and providing for modular growth when the system is subject to future requirements.

Depending upon the type of equipment, these processes can be complex. As a result, the telephone industry has spent considerable time in the development of traffic engineering practices and tables. A large portion of this book is devoted to the presentation of methods and explanation of the reasoning behind those methods. Traffic tables provide good reference material; however, sound engineering judgment is essential in their application. We must develop a better understanding of traffic system concepts in order to exercise better judgment for resolving conflicting requirements and achieving a cost-effective system design.

1.1.3 Traffic Order

The traffic order (TO) is the specification that describes the results of the traffic system design process. It defines the configuration of equipment, specifies equipment quantities, and provides general information related to the operation of the planned facility. There is another important function that the TO fulfills—it must justify the capital investment. This is important for budget control and financial integrity. See section 7.1.2 for a discussion of the role the TO plays in the system planning process.

1.2 EVOLUTION OF SWITCHING

Bell's patent for the telephone was approved in 1876. In the early days of telephony, a telephone hookup was simple—wires connected pairs of telephones, and communication was possible only with the telephone at the other end. As the instruments improved, and more people became interested in telephone service, the necessity for rapidly interconnecting telephones became apparent.

The most obvious method was to connect every telephone to every other telephone. But this procedure quickly became very complicated. Fifteen connections are required to connect only six telephones in this manner as depicted in Figure 1.1. Over twelve hundred different connections would be required to interconnect only fifty telephones.

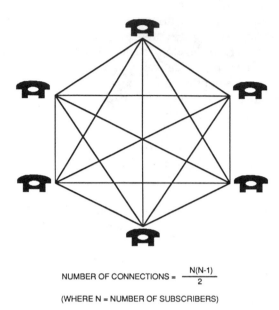

NUMBER OF CONNECTIONS = $\dfrac{N(N-1)}{2}$

(WHERE N = NUMBER OF SUBSCRIBERS)

Figure 1.1 Direct Connection of Telephones

The logical and sensible way to solve the problem of telephone interconnection was to gather up the wires and bring them together at a central switching point as depicted in Figure 1.2. Then all that would be required would be a means of connecting the right wires together when subscribers wanted to talk to each other.

1.2.1 Manual Switchboards

Initially, the telephone was looked upon as a luxury item for professional people to keep in touch with their business associates. Because of this feeling, it was usually possible to serve all of the customers in a town from a single manual switchboard. Connections were made by extending one line to another by means of a line cord between jacks on a board. As the telephone ceased to be considered a luxury and as one central office (CO), because of its physical limitations, could no longer serve the demands of a city, additional offices were set up. Interoffice trunks were then used to permit access from one office to the other, and a second operator was required to complete the call.

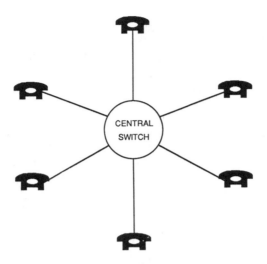

NUMBER OF CONNECTIONS = NUMBER OF SUBSCRIBERS

Figure 1.2 Switched Connection of Telephones

As local calling grew, so did long-distance calling. Cities soon required special switchboards to handle inward calls and other switchboards for through calls. A long-distance call was built up as an operator in one city passed the call to an operator in a second city, and so on, before reaching the destination switchboard and completing the connection. This method was slow in operation and expensive in operator work time.

1.2.2 Automatic Dial Operation

Local dial offices came into use with the introduction of the Strowger stepper switch in 1892. Dialing provided the greater speed, accuracy, and economy that manual operation lacked. Direct-distance dialing (DDD) came into being for the same reasons operator-distance dialing came into being. By prefixing certain codes, operators could automatically obtain a circuit to a distant city via a tandem office and establish a connection to the called telephone. Inward boards remained only at manual offices to provide assistance elsewhere.

An integrated complex of these electromechanical step-by-step (S × S) switching machines and intercity circuits grew to become a network over which operators could be rapidly and automatically connected to a desired telephone. In an area with a large number of subscribers, a limitation had to be placed on the number of subscribers in one group of switchboards. It was decided to adopt a basic four-digit numbering plan so that each CO would serve up to 10,000 subscriber lines.

1.2.3 Common-Control Systems

Common-control switching is as old as telephony. The earliest form was the manual switchboard with the operator fulfilling the role of the control element. The lamps and position circuit represented interface equipment, and the cord circuit established the transmission path. Years of growth in subscribers and sophistication led the industry back to the same basic principle, and technological improvements accelerated this trend in recent years. Although there are still manual- and progressive-control systems in operation, they are rapidly being phased out and replaced by stored-program control systems.

This book focuses on common-control switching systems. The distinguishing characteristic of common-control switching systems is that the switching logic is separate from the switching matrix, which is a switching device with no logical capacity. The matrix is acted upon by interface equipment, as directed by the common control.

The principal elements of a common-control system are the switching matrix, control equipment, and interface equipment. The switching matrix provides the transmission path through the switch, the control equipment directs machine activity, and the interface equipment facilitates communication between the other two elements. The allocation of these basic functions is presented in Table 1.1.

The two methods used to implement switching matrices in common-control systems are stage-by-stage selection and conjugate-link selection. Stage-by-stage link selection is not to be confused with S × S switching, although the latter is a subset of the former. *Stage-by-stage link selection* implies that the total transmission path is not established in one operation. Typically, such a system will employ a line matrix and a trunk matrix which are operated upon by different control units in succession. This tends to minimize sophistication in control and memories, but suffers from the fact that the state of the entire matrix is not considered in establishing the connections. Conversely, *conjugate-link selection* implies the consideration

Table 1.1 Allocation of Switching Functions

Function	Matrix	Control	Interface
Interconnecting	X		
Controlling		X	
Alerting			X
Attending			X
Receiving			X
Transmitting			X
Busy Testing			X
Supervising			X

of the entire matrix in establishing the connections between the inlet and outlet terminations. This makes the most efficient use of the matrix, but requires more time and memory to perform matrix operations.

1.3 NUMBERING PLANS

An essential element of any telephone network is a numbering plan wherein each subscriber has a unique directory number which is convenient to use, readily understood, and similar in form to that of all other subscribers connected to the network. This number, when dialed from any location, acts as a destination code. It provides all the routing information required for the switching systems involved in establishing the connection.

The ten-digit numbering plan established for the North American DDD system consists of a three-digit numbering plan area (NPA) code, a three-digit office code, and a four-digit station code. Typical subscriber dialing procedures are given in Table 1.2. Note that one or zero in the first digit position are reserved for operator, toll, and other special codes.

Table 1.2 North American Dialing Procedures

Type of Call		Dialed Digits
Local		NNX-XXXX
Station Toll	- HNPA	1-NNX-XXXX
	- FNPA	1-NYX-NNX-XXXX
Special Toll	- HNPA	0-NNX-XXXX
	- FNPA	0-NYX-NNX-XXXX
WATS (Toll-Free)		1-800-NNX-XXXX
Directory Assistance	- Local	411
	- HNPA Toll	1-555-1212
	- FNPA Toll	1-NYX-555-1212
	- WATS	1-800-555-1212
Other Service Codes		N11 or NNX-XXXX
Operator Assistance		0
International Access		01 or 011
Special Codes		00, 10 and 11

where

> N = any digit 2 through 9
> X = any digit 0 through 9
> Y = either 0 or 1
> HNPA = home (local) area code
> FNPA = foreign (other) area code
> WATS = wide area telecommunication service

1.3.1 Numbering Plan Areas

North American NPA boundaries are drawn coincident with state and province (Canadian) boundaries when the ultimate number of CO codes required does not exceed the code capacity of an NPA. A single NPA for a state or province is the most desirable arrangement from a subscriber dialing viewpoint. The boundary is familiar and only one area code is needed to dial any telephone within the state or province. In the more populous states this is not possible. In these cases two or more area codes are used depending on the number of offices to be served.

Area codes consist of three digits (NYX) wherein the first digit (N) may be any digit 2 through 9 and, at present, the second digit (Y) is always either 0 or 1. With this format, switching equipment can distinguish area codes from office codes because the latter normally do not use 0 or 1 for the second digit (there are exceptions, however).

There are 160 possible NYX combinations which are now available for use as area codes. Of these, eight N11 codes (211, 311, . . . , 911) are reserved for use as service codes, leaving 152 available for use as area codes. When the spare NYX codes are exhausted, additional assignments will be made from NN0 codes. Then the remaining unused NXX codes will be used for area codes.

A local access and transport area (LATA) is defined as the geographic area within which the telephone company provides all local and long-distance services, plus access to the PSTN. Long-distance traffic between LATAs (inter-LATA calls) is handled by the common carriers (AT&T, MCI, US Sprint, *et cetera*). In general, LATA boundaries coincide with NPA boundaries, but here also there are exceptions.

In subdividing states, boundaries are generally drawn such that points generating high volumes of traffic to each other are in the same LATA. This permits the customer to dial only seven digits for these calls. Wherever possible, the boundaries are not drawn between a toll center and its tributary offices. As additional area codes are introduced, the same principles will apply in determining the new boundaries. The controlling objectives are to keep the numbering plan as simple as possible, and to enable the subscriber to complete as many calls as possible by dialing only seven digits.

1.3.2 Office Codes

The theoretical quantity of CO codes of the formerly used two-letter plus five-digit (2L + 5D) form is 640, because the letters on the dial only appear in the 2 through 9 positions. The all-number CO codes (NNX) increase the usable codes to 780. Of the 800 NXX combinations, 202 are currently reserved for area codes, service codes, and special purpose codes.

1.3.3 Access Codes

The standard procedure for subscriber-dialed toll calls is to prefix the digit 1 to the called number. This enables calls to be switched into the toll network in S × S offices, and is also used to distinguish between toll and local calls. The prefix digit 0 is used for subscriber access on operator-assisted calls such as person-to-person, collect, and credit-card calls. International access codes 01 and 011 enable subscribers to directly dial other subscribers around the world.

Private telephone systems also employ access codes for special purposes. For example, PABX (private automatic branch exchange) switches commonly use the prefix digit 9 for access to the local CO, and may use the digit 8 (or some other convenient digit or digits) for access to a private corporate dial network. In such systems, the access digit selected may not be used as the first digit of any PABX extension. This reduces the number of possible extension codes available on the PABX. However, this is not a problem because there are normally a limited number of subscribers to be served.

1.4 SIGNALING PLANS

Signaling is the means employed to transfer call information in a telecommunication network. *Line signaling*, also known as supervision, is used to control the setup, holding, charging, and releasing of call connections. The primary line signaling functions are to initiate a request for service, indicate that digits may be sent, alert the called terminal (station), indicate that the called party has answered, relay call-charging information, and indicate when the call is completed. *Register signaling*, also known as address signaling, is used to transmit the called subscriber's address (directory number) through the network to the terminating switch.

Audible tones and recorded announcements are used to provide call progress indications to the calling party, such as, digits may be sent (dial tone); called station is being rung (ringback tone); called station is busy (busy tone); call cannot be completed due to network congestion (reorder tone); and dialing error (error tone).

Signaling may also be classified into the general categories of subscriber line signaling and interoffice trunk signaling as depicted in Figure 1.3. *Subscriber line signaling* includes all signaling between a subscriber terminal and a switch. *Interoffice trunk signaling* includes all signaling between switches to manage the call connection through the network. Interoffice trunk signaling is analogous to subscriber line signaling, except that the signals are machine-initiated for greater speed and accuracy.

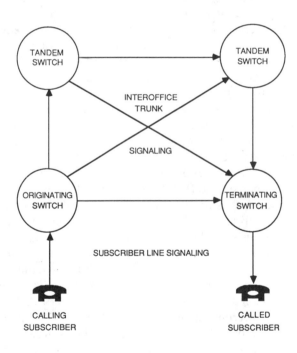

Figure 1.3 Signaling Category Diagram

1.4.1 Subscriber Line Signaling

Subscriber line-signaling techniques vary depending on the type of service provided. An important concept to understand is that the speed of dialing, hence the speed of setting up a call, is in great measure dictated by the type of customer service.

In the early days of manual ringdown (RD) signaling, the calling subscriber had to send a ringing signal to the local office to initiate a call. The switch sent all incoming calls to operators, who asked the calling subscriber for the called subscriber's name or number, and then extended the call manually with line cords at the switchboard. An operator at the terminating switch rang the called subscriber's telephone, while the operator at the originating switch listened for the called subscriber to answer so that the call could be properly charged (billed). The subscribers had to ring off at the end of the call to notify the operators that the call was completed (stop charging). Obviously, this was an inefficient use of subscriber, operator, and facility time.

Common-battery supervision (CBS) was an improvement on RD service in that a flow of electrical current in the subscriber loop alerted the CO of the on-hook and off-hook status of the subscribers. This eliminated the hand crank, but was still a labor-intensive method of signaling—all calls still had to be handled by operators.

The stepper switch enabled subscribers to dial calls without operator assistance. When the CO detects a subscriber off-hook status and is ready to accept digits, it sends a dial tone to the calling subscriber. The calling subscriber then dials the called subscriber's directory number. The switch detects dialed digits as a series of on-hook and off-hook pulses, at a nominal rate of ten pulses per second (PPS), on the line. The stepper switches step once for each pulse, thus routing the call to the called subscriber's line or outgoing trunk, as applicable. Called *rotary-dial* or *dial-pulse* (DP) service, it is still used extensively throughout the world, although more modern switching equipment, such as crossbar switches, may be employed.

Dual-tone multifrequency (DTMF) pushbutton signaling uses unique pairs of frequencies to represent digits. A distinct advantage of DTMF signaling over DP signaling lies in the relative speed of dialing. Table 1.3 presents typical subscriber dialing times for DP and DTMF signaling as a function of the number of digits to be dialed. Dial-pulse signaling speed is limited by the inherent ten PPS speed of the rotary dial, whereas DTMF signaling speed is limited essentially by the subscriber's dexterity (or the speed of the increasingly popular automatic dialers).

Table 1.3 Typical Subscriber Dialing Times

(Time in Seconds)

Type	Number of Digits Dialed				
	1	4	7	10	11
DP	3.7	8.3	12.8	17.6	19.1
DTMF	2.3	5.2	8.1	11.0	12.0

1.4.2 E&M Trunk Signaling

Most modern interoffice trunks use four wires—a transmit pair and a receive pair. In addition, E&M* trunks use two or three more wires for line signaling (supervision). In the idle state, the M-leads at each end of the trunk are grounded so that no current flows in the supervision loops. The calling (originating) switch sends a seize signal by closing its M-lead to battery. The called switch detects the resultant current flow at its E-lead, and closes then reopens (winks) its M-lead to acknowledge the seizure and indicate its readiness to receive digits. The calling switch then sends DP digits by pulsing the M-lead. The called switch detects the pulses at its E-lead, translates the digits, and routes the call on toward the called subscriber's office (terminating switch).

When the called subscriber answers, the terminating switch sends an answer signal back toward the originating switch by closing the M-lead to battery continuously. Both M-leads remain closed to battery for the duration of the call. An E&M trunk circuit is released by one switch grounding the M-lead continuously, and the other switch acknowledging the release by grounding its M-lead continuously.

Obvious disadvantages of E&M signaling include the additional supervision wires, signaling power requirements, and the inability to multiplex and transmit the dc-closure E&M signals over radio or optical transmission links.

---

*From the Latin words *excipere* (to receive) and *mittere* (to send), more easily remembered as "ear" and "mouth."

1.4.3 Single-Frequency Trunk Signaling

Four-wire, single-frequency (SF) signaling trunks overcome the disadvantages inherent in E&M signaling. A 2600 Hz tone is sent both ways over the trunk to indicate the idle condition (tone-on-idle). Signaling on SF trunks is analogous to E&M signaling where the presence of the tone has the same meaning as grounding an M-lead, and the absence of the tone has the same meaning as closing an M-lead to battery. The SF tone, being within the voice band, can be multiplexed for transmission over radio or optical links. Signaling converters are available to convert E&M signals to SF signals and *vice versa*.

1.4.4 Multifrequency Trunk Signaling

Four-wire, multifrequency (MF) signaling is also referred to as MF 2/6 signaling because it uses unique pairs of six tones. The tone pairs are different than those used for DTMF, and the signaling speed is much faster. Multifrequency signaling is commonly used for intertoll trunks and international gateway exchanges (CCITT #5), and foreign national signaling (CCITT R1 and R2). CCITT R2 signaling is often referred to as MFC because it is compelled signaling. That is, for each MF tone pair there is a complementary tone pair which is sent back to the calling switch to verify that the correct signal was received at the called switch. Therefore, MFC signaling requires about twice the time of MF signaling, but has the advantage of assuring the calling switch that the called switch received the transmitted signal correctly. Table 1.4 presents typical MF signaling times as a function of the number of digits sent.

Table 1.4 Typical Multifrequency Signaling Times

(Time in Seconds)

Type	Number of Digits Sent				
	1	4	7	10	11
Sender	1.5	1.9	2.3	2.8	3.0
Receiver	1.0	1.4	1.8	2.2	2.3

1.4.5 Common-Channel Interoffice Signaling

Modern switching systems equipped with electronic switches typically use common-channel interoffice signaling (CCIS). All of the signaling information pertaining to a trunk group is transmitted in digital messages over a dedicated signaling channel, and traffic is transmitted over the other channels. The main advantage of CCIS is the reduction in call-setup time because all signaling, including network management messages, takes place between the call processors over the dedicated signaling channel. Other advantages are signaling flexibility and low cost for interoffice links with a high volume of signaling. One of the disadvantages of CCIS is that the entire link would be disabled if the signaling channel failed. This may be avoided or minimized by duplicating the signaling channel for higher reliability.

1.4.6 Typical Signaling Sequence

Consider a typical subscriber-dialed call as depicted in Figure 1.4. The calling subscriber goes off-hook to initiate the call. The originating CO detects the off-hook, assigns an appropriate signaling register (receiver) to the line, and sends a dial tone to the subscriber. The calling subscriber then dials the called subscriber's directory number (address), and the receiver collects the digits. When all the digits are received and transferred to the digit translation process, the receiver is released and becomes available to serve another call. The receiver-holding time is therefore the duration of the signaling phase, and is a function of the number of digits dialed, the dialing technique (DP or DTMF), and the subscriber's dialing speed.

Assuming the call is to a subscriber at an adjacent CO, the originating CO selects an outgoing trunk to the terminating CO and institutes trunk signaling. This involves assigning an outgoing address signaling register (sender) and sending a seize signal on the trunk. When the terminating CO detects the seize signal, it assigns an incoming trunk signaling register (receiver) and sends a start-dial signal to the originating CO. The originating CO then sends the address digits to the terminating CO, releases the outgoing sender, and connects the calling subscriber's termination to the outgoing trunk termination. The terminating CO releases the incoming receiver when all the address digits have been received. The register-holding times are the duration of the trunk signaling phase, and are a function of the number of digits sent and the trunk signaling technique employed.

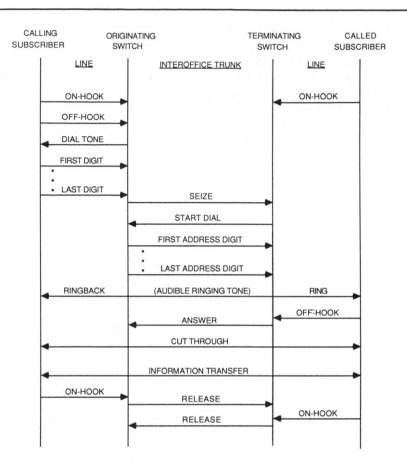

Figure 1.4 Typical Signaling Sequence

If the called subscriber is busy (off-hook), the terminating CO sends a busy tone to the calling subscriber over the connection. If the called subscriber is idle (on-hook), the terminating CO rings the called subscriber's termination and sends an audible ringing tone (ringback) to the calling subscriber. When the called subscriber answers, the terminating CO sends an answer signal to the originating CO and connects the called subscriber's termination to the incoming trunk termination. This is commonly called *cut through*, and the subscribers may transfer information to each other.

Either subscriber may initiate release by hanging up (go on-hook). The local CO detects the on-hook status and sends a release signal over the trunk to the other CO. After the exchange of release signals, the trunk is idle (disconnected) and available to handle a subsequent call. The holding time for the trunk circuit is the duration from seizure to release. The billable call-holding time is the time from answer signal until release. Thus, it can be seen that the signaling time must be minimized in order to maximize the efficiency of network facilities.

1.5   ROUTING PLANS

Each day there are tens of millions of toll calls generated over the PSTN. This traffic is routed over hundreds of thousands of intertoll trunks and is switched by thousands of toll offices. To handle this volume of traffic, it was necessary to create a hierarchical routing plan that classifies offices according to function, and to route traffic according to the class assigned to each office. However, today's post-divestiture network, with competing long-distance carriers, can hardly be considered hierarchical.

The routing pattern for a call between any two points is determined by the final routing path between the originating and terminating offices. A call may switch only at switching centers on the final route. It may be offered to any high-usage trunk group that bypasses one or more switching centers in the final chain, as long as the call always progresses toward its destination. Calls between high-volume points are normally completed on direct trunks regardless of distance. As traffic volume increases, a larger portion of the traffic is carried on direct routes.

When the volume of traffic between two offices is small, however, the use of direct trunks is usually not economical. In these cases, the traffic is handled by interconnection via switching equipment at intermediate offices to build up the required connection. The switching centers where interconnections are made are generally referred to as *tandem switches*. Built-up connections may involve several tandem switches if the originating and terminating locations are a great distance apart. The resultant signaling between each pair of switches will significantly increase the call-setup time, thus lowering the efficiency of the network facilities used to complete the call.

1.5.1   Alternate Routing

The needs of distance dialing are met by switching and trunking arrangements that use alternate routing. Alternate routing is a deterministic arrangement in a switching center whereby a call is offered to another circuit group when all circuits in the primary group are busy. Figure 1.5 depicts typical single-stage alternate routing. Calls to a distant point may be offered successively to several trunk groups prior to being offered to the final group. Figure 1.6 depicts typical multistage alternate routing.

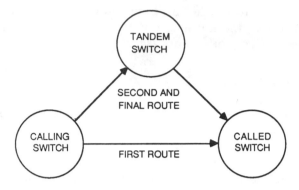

Figure 1.5 Single-Stage Alternate Routing

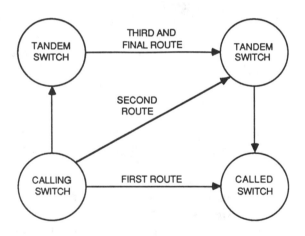

Figure 1.6 Multistage Alternate Routing

1.5.2   Intertoll Trunks

Intertoll trunks are provided between offices where the traffic volume warrants it. Usually, these trunks are two way; that is, they may be used for calls originating at either end. There are exceptions in very large groups where it is often more economical to have some groups arranged for one-way operation. Intertoll trunks are classified by function, as follows:

a. High-usage trunk groups are not engineered to carry all of the traffic between two points, thus offering overflow traffic to an alternate trunk group, which may be another high-usage group or a final group.
b. Final trunk groups do not have an alternate trunk group to handle their overflow. Calls that cannot obtain a circuit in the final group fail to complete. Therefore, final groups are normally engineered on the basis of one call per hundred (1% or P.01) blocked during the busy hour of the busy season.

1.5.3 Adaptive Routing

Adaptive routing, also known as saturation or flood-search routing, is a nondeterministic extension of alternate routing. The network is flooded with messages that search for a fixed-directory number without regard to where that number is terminated in the network (Ludwig and Roy 1977). Adaptive routing is presently employed in some tactical military networks where it is imperative to have fast, reliable access to mobile subscribers anywhere within the network. These networks employ digital switching systems using CCIS to handle the high volume of signaling traffic.

1.6 DATA SERVICES

The PSTN is handling ever-increasing data-traffic loads. Data traffic carried in voice-band facilities includes terminal to central-computer traffic, central computer to terminal traffic, and time-share operations. Packet switches and local-area networks (LAN) typically handle traffic between computers, multiple users to computers, and computer database access. The PSTN currently supports data traffic up to 2400 bits per second (b/s) with relatively inexpensive modems. More expensive modems are employed to provide higher data rates.

Data traffic characteristics are quite different from those of voice traffic. Therefore, a word of caution is in order when attempting to use formulas and tables based on traffic data collected from long experience with the switched voice network. The characteristics of the specific system must be compared with the underlying assumptions of the formulas and tables to determine if they are appropriate for that application.

1.6.1 Packet Switching

Packet switching systems dynamically switch and transmit data throughout the network. Data packets from all users are multiplexed on common (shared) digital network resources, imposing increased traffic requirements. A number of packet transmission protocols, such as CCITT X.25, have been developed for these applications.

Bandwidth requirements for the different users may vary significantly over time. The network has been designed to tolerate moderate variations, but if peak traffic loads exceed the limits then network facilities, such as switches and transmission media, can be affected. New network architectures are being developed to improve system performance control for packet switching (Zhang 1987).

1.6.2 Local-Area Networks

In contrast to circuit- and packet-switching networks, LANs are intended to be point-to-point or point-to-multipoint high-speed networks. Each station (user) recognizes information destined for it by a unique address in the message header. A LAN is usually implemented as a closed system operating in a specific geographical area; however, it may connect into a packet-switching network at a gateway exchange. A number of LAN protocols, such as IEEE 802.3, have been developed for these applications.

A typical LAN employs a bus in a linear, ring, or star network. A *linear bus network* is a length of transmission medium, such as a coaxial cable, with station taps along its length for the connection of users. A *ring system* attaches all stations to a functionally circular bus. Messages are passed around the ring in one direction only, and each station passes on the messages not destined for it. *Star networks* are hierarchical, similar to the switched voice network, in that they employ central nodes with stations terminating on them.

1.6.3 Integrated Services Digital Network

The constantly increasing volume of data traffic on the PSTN is forcing the evolution of the integrated services digital network (ISDN) to consolidate circuit, message, and packet switching over existing facilities. Many ISDN implementations are currently based on performance and technical standards that are unique to the individual manufacturers. The result of this approach is that subscribers do not have universal access to all other subscribers. Another problem with ISDN is the current high cost—a sufficient number of subscribers must be available to achieve the economies of scale necessary to lower the cost to a level that will attract more subscribers. When the full potential of ISDN is realized, it will have a significant effect on the traffic carried by the PSTN.

REFERENCES

Bellamy, J., *Digital Telephony*, New York: John Wiley and Sons, 1982.

Bell Laboratories, *Engineering and Operations in the Bell System*, 2nd Ed., Indianapolis: Western Electric, 1983.

Ludwig, G., and R. Roy, "Saturation Routing Network Limits," *Proceedings of the IEEE*, Vol. 65, No. 9, September 1977, pp. 1353–1362.

Martin, J., *Future Developments in Telecommunications*, 2nd Ed., Englewood Cliffs, NJ: Prentice-Hall, 1977.

Zhang, L., "Designing a New Architecture for Packet Switching Communication Networks," *IEEE Communications Magazine*, Vol. 25, No. 9, September 1987, pp. 5–12.

Chapter 2

Traffic Characteristics and Service Objectives

This chapter presents the principles of teletraffic engineering theory, and the assumptions that underlie traffic formulas and tables. The basic factors involved in solving teletraffic engineering problems are the volume and nature of traffic to be handled, the facilities to serve the demand, and the grade of service desired.

The traffic problem has three sides, similar to a triangle. Whenever the dimensions of one side are changed, one (or both) of the remaining two sides must also change. If, for instance, the demand changes and the facilities remain the same, the grade of service will change. If we wish to change the grade of service, the demand or the facilities must be changed. A highway may be used as a familiar analogy to illustrate this concept. If the number of lanes (facilities) in the highway remains constant, increases in the number of vehicles using the highway (traffic flow) will increase the level of congestion (grade of service). Similarly, if traffic flow is relatively constant, but an accident shuts down a lane, then congestion also increases.

2.1 TRAFFIC TERMINOLOGY

When dealing with telephone traffic in general, and traffic units in particular, certain terms should be defined and understood at the outset.

2.1.1 Call, Call Attempt, and Call-Attempt Factor

A *call attempt* is defined as any effort on the part of a traffic source (subscriber) to obtain service. A *call* is a series of dialing attempts to the same number, where the last attempt is either abandoned or results in a successful call. Thus, the number of successful calls is equal to or less than the number of calls, and the number of calls is equal to or less than the number of call attempts.

The *call-attempt factor* (CAF) reflects the fact that not all call attempts become completed calls—a typical completion rate for DDD voice calls is 70 percent, leading to a CAF of about 1.4. A call may be cut short because the called subscriber's line is busy, an all-trunks-busy (ATB) condition is encountered, the calling subscriber misdialed and had to try again, *et cetera.* All of these call attempts demand signaling register service, and therefore contribute to server traffic. They also tend to decrease the mean call-holding time.

2.1.2 Call-Holding Time

Call-holding time is the length of time during which a traffic source engages a traffic path or channel. Frequency distributions of typical voice call-holding times yield charts similar to that of Figure 2.1. Call-holding times of one to three minutes are relatively frequent, whereas call-holding times longer than ten minutes are more infrequent. A common assumption in telephone traffic theory is that call-holding times have a negative-exponential tendency, as indicated by the curve superimposed on the frequency distribution. Experience over many years has shown that use of the negative-exponential distribution for voice calls is justified. However, holding times for data calls may range from many short calls (e.g., packet switching) to a few very long calls (e.g., computer data dumps). Therefore, these call-holding times may be relatively constant, and use of the negative-exponential assumption would not be justified.

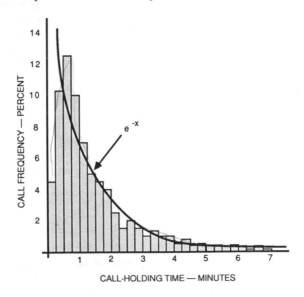

Figure 2.1 Typical Voice-Call Holding-Time Distribution

2.1.3 Busy Hour

The *busy hour* is that continuous sixty-minute period of the day during which the highest usage occurs. It does not have to coincide with the clock hour; however, in order to simplify traffic measurement, the busy hour usually commences on the hour, half hour, or quarter hour. When busy-hour traffic data are used, we should know whether the data reflect the busiest hour of an entire busy season, month, or week, or if they reflect the busiest hour of the busiest day of a busy season, month, or week.

The method of determining the representative busy-hour traffic intensity for computing quantities of hardware varies from application to application. However, the mathematical formulas used to determine the quantity of switches are based on the assumption that the busy-hour intensity is the average of an infinite number of busy hours.

2.1.4 Traffic Density and Traffic Intensity

Traffic density is defined as the number of simultaneous calls at a given moment, whereas *traffic intensity* represents the average traffic density (occupancy) during a one-hour period. Occupancy is any use of a traffic resource, regardless of whether or not a connection (call) is completed. The traffic figure employed in traffic capacity tables is the busy-hour traffic density.

2.1.5 Offered, Carried, and Blocked Traffic

In loss systems, *offered traffic* is the traffic intensity that would occur if all traffic submitted to a group of circuits could be processed by the group. *Carried traffic* is the traffic intensity actually handled by the group. The traffic-carrying capacity is the nominal value of traffic carried, with a given loss probability, by the group. *Blocked traffic* is that portion of traffic that cannot be processed by the group (offered traffic minus carried traffic). Blocked traffic may be rejected, retried, or offered to another group. Blocked traffic offered to another group is commonly called overflow traffic.

2.1.6 Loss and Delay Systems

A *loss system* is a system in which a call attempt is rejected when an idle resource is unavailable to serve the call. A *delay system* is a system in which call attempts are held in a waiting queue until resources are available to serve the calls.

2.2 TRAFFIC UNITS

Telephone traffic may be defined as the occupancy of the transmission and switching facilities that comprise the network during the process of establishing a connection and while the call is in progress. It is important to emphasize that traffic is generated from the moment the calling subscriber lifts the handset (goes off-hook), until the call is released by replacing the handset back on its cradle (goes on-hook).

2.2.1 Traffic Flow

Traffic flow through a switching office is defined as the product of the number of calls during a period of time and their average duration. Traffic flow, therefore, can be expressed by the equation:

Traffic Flow = (Number of Calls)(Mean Call-Holding Time) (2.1)

For example, if 100 calls of an average duration of three minutes are generated during a period of one hour by the subscribers connected to a line-finder group, then the traffic flow for the group equals 300 call-minutes, or 5 call-hours.

2.2.2 Erlangs and Unit Calls

The international, dimensionless unit of telephone traffic is called the Erlang, named after A.K. Erlang, the father of telephone traffic theory. One Erlang represents a circuit occupied for one hour; thus,

$$1 \text{ Erlang} = 1 \text{ Call-Hour/Hour} \tag{2.2}$$

The number of Erlangs per busy hour may be calculated as follows (call-holding time expressed in hours):

$$\text{Erlangs} = (\text{Calls/Busy Hour})(\text{Mean Call-Holding Time}) \tag{2.3}$$

For example, consider a connection established at 2:00 AM between a central computer and a data terminal. Assuming that the connection was maintained continuously and data was transferred at a rate of 1200 b/s, determine the amount of traffic in Erlangs transferred over the established connection between 2:00 AM and 2:45 AM:

$$\text{Traffic} = (1 \text{ call})(45 \text{ min})(1 \text{ h}/60 \text{ min}) = 0.75 \text{ Erlangs}$$

Note that the data rate is immaterial—the traffic is based solely on the call-holding time which, in this case, is 45 minutes or 0.75 hours.

When the call-holding time is expressed in seconds, the resulting traffic unit is the *unit call* (UC), or its synonymous terms *hundred-call-seconds* or *centum-call-seconds* (CCS), as expressed in (2.4):

CCS = (Calls/Busy Hour)(Mean Call-Holding Time)/100   (2.4)

For example, consider a 5000-line office that has a busy-hour calling rate of 0.6 and an average call-holding time of 250 seconds. What is the busy-hour traffic in CCS?

Traffic = (5000 lines)(0.6 calls/line)(250 s/call)/100 = 7500 CCS

Because there are 3600 seconds in an hour, the relationship between Erlangs and CCS is:

1 Erlang = 36 CCS   (2.5)

For example, consider that a group of 20 subscribers originate a total of 50 calls with an average holding time of 200 seconds during the busy hour. How many CCS and how many Erlangs per subscriber is this?

Traffic = (50 calls)(200 s/call)/100 = 100 CCS

100 CCS/20 subscribers = 5 CCS per subscriber

5 CCS/36 ≈ 0.139 Erlangs per subscriber

Even though CCS is the traffic unit employed predominantly in North America, we should recognize that traffic expressed in Erlangs is more meaningful and provides more direct information than traffic expressed in CCS. Consider the following:

a. The Erlang per channel represents its efficiency; that is, the proportion of the hour during which the channel is occupied.
b. Traffic expressed in Erlangs designates the average number of calls in progress simultaneously during a period of one hour.
c. Erlang figures represent the total time, expressed in hours, to carry all calls.

For example, if a group of 30 trunks are required to carry 600 CCS of traffic during the busy hour, what is the efficiency of this trunk group?

600 CCS/36 ≈ 16.7 Erlangs

16.7 Erlangs/30 trunks $\approx$ 0.557 Erlangs per trunk

Trunk-group efficiency $\approx$ 0.557 = 55.7 percent

You will find that the Erlang is used exclusively as the traffic unit in classical traffic theory. Familiarity with both units is essential because telephone systems designed for international and military applications commonly provide traffic data in Erlangs. As the United States resists conversion to metric units of measurement, so it persists in using CCS to measure telephone traffic intensity.

2.3 VOLUME AND NATURE OF TRAFFIC

An understanding of the volume and nature of traffic and its distribution with respect to time is essential in determining the facilities required to serve the system's needs. This is particularly true in cases where common-control equipment is used in switching traffic. Absolute figures of traffic volume are not sufficient for sound traffic engineering.

2.3.1 Variations in Traffic Intensity

Telephone traffic varies greatly from one period to another, not in any uniform manner, but according to the needs and whims of the subscribers. Traffic volumes vary from season to season, month to month, day to day, and hour to hour. There are also variations from minute to minute in the same hour. Traffic volume within the week usually forms a fairly consistent pattern, such as high on Mondays and Fridays, lower on Wednesdays and lowest on Sundays. Other factors, such as holidays and weather, can affect the weekly traffic pattern. Although the traffic volume during business hours is significant, it is the busy-hour traffic that is of most interest to the traffic engineer.

Traffic-sensitive equipment requirements are determined on the basis of the traffic intensity, which is based on normal busy seasons. Occasionally, however, the traffic will exceed that level. Some peaks can be expected on such days as Mother's Day and Christmas, while other peaks are unexpected. To engineer the system for all such contingencies would be very costly. The broad guidelines in using traffic data are as follows:

a. use traffic data pertaining to four consecutive high weeks for engineering CO equipment and trunks;
b. use traffic data pertaining to ten high days for engineering common-control equipment.

2.3.2 Service Objectives

The other set of design parameters is concerned with service objectives. The term *service objective* can be used in several different ways to determine an optimal level of service. Within the context of teletraffic engineering, the terms *service objective* and *grade of service* are used synonymously to define a level of traffic congestion.

There is some ambiguity as to what is considered optimum. This difficulty stems partly from the fact that the cost *versus* service curve is a continuously increasing function. It is also due to the subjective nature of the subscriber's evaluation, which is unavoidably a composite view of many factors (some of which are totally unrelated). Thus, what represents a good level of service to one subscriber may be intolerable to the next. In order to quantitatively define the quality of service, it is necessary to precisely identify the criteria for that measurement. These are the degree of congestion and the basis for that measurement.

The *degree of congestion* defines the frequency with which a subscriber will encounter some measure of inconvenience in completing a call, either because it has to be repeated or because it must wait for service. The degree of congestion, then, can be expressed mathematically as the probability that an unfavorable event will occur, or as the average delay that can be expected, under specific traffic loads, characteristics, and arrangements.

Establishing a *basis* from which the degree of congestion is measured is necessary simply because traffic volumes are not constant, but fluctuate. As a result, the traffic load used in estimating the relative degree of congestion must represent the usage in a specific time interval, or the accumulated results of several such study intervals. Therefore, the service objective must not only define the probability of congestion, but must indicate the frame of reference.

2.4   GRADE OF SERVICE

Grade of service is a measure of the probability that a percentage of the offered traffic will be blocked or delayed. Grade of service, therefore, involves not only the ability of a system to interconnect subscribers, but the rapidity with which the interconnections are made. As such, grade of service is commonly expressed as the fraction of calls or demands that fail to receive immediate service (blocked calls), or the fraction of calls that are forced to wait longer than a given time for service (delayed calls).

The assessment of the grade of service provided to the user, and the determination of facilities required to provide a desirable grade of service, are based on mathematical formulas derived from statistics and the laws of probability. An understanding of the fundamental principles of statistics and probability is essential to an understanding of traffic distributions and formulas (see Section 2.6).

2.4.1   Blocking Criteria

If a system is engineered on the basis of the fraction of calls blocked, then it is said to be engineered on a blocking basis. Blocking can occur if all facilities are occupied when a demand is originated or, in the case where several facilities must simultaneously be connected, a matching of idle facilities cannot be made even though certain facilities are idle in each group. Some areas where engineering is done on the basis of blocking criteria are the dimensioning of switching matrices and interoffice trunk groups.

2.4.2   Delay Criteria

The fraction of calls delayed longer than an acceptable time provides another basis for setting a service standard. Such a standard would state "the fraction of calls delayed longer than a given time shall not exceed a given value," and the system would be engineered to meet this criterion. The delay encountered in providing a dial tone through an idle register to a subscriber is an example. As the number of common-control and time-sharing systems increases, so does the need for engineering on a delay basis.

2.4.3   Blocking and Delay Formulas

Blocking and delay formulas deal with sources (subscribers) offering traffic to a group of servers (operators or equipment). These formulas were derived based on a series of assumptions (see Chapter 3). As the load is increased, other parameters remaining the same, the grade of service will deteriorate. Deterioration to the point of just meeting a service objective is economically desirable, but beyond that point it is undesirable or un-acceptable. In some cases it may have a detrimental effect on other parts of the system. For example, a shortage of incoming signaling registers (receivers) in a CO will result in slowing down of the originating office senders. The sender-holding time is increased because it includes the time the switch must wait for the receivers at the distant office (there is a good deal of such interdependence in every switching system). When server blockage or delay is kept within bounds, most of these interdependence effects are small and can usually be ignored. However, in the above example, the traffic engineer at the originating office might request additional senders when, in fact, the number of terminal receivers should be increased.

There are other parts of the system where dependence should not be ignored, either because economics dictates that service levels of the individual groups of servers cannot be so low as to eliminate any need to consider interdependence, or where extreme peakedness exists. During peak traffic periods, interdependence may make the theoretical service results unrealistic. Some examples of this are reflected in dial-tone delay in S × S offices where dependence exists between the line finders and the subscriber dial-tone senders; in crossbar systems between the dial-tone markers and originating registers; and in call-distribution systems that require calls to be channeled one at a time through a distribution circuit to gain access to a team of operators.

2.5 QUEUING SYSTEMS

When a system is operating on a delay basis and the congestion point is reached, any new call arrival must wait until a server is available. The new call, or a prior call if others are waiting, will then be served. While these calls are waiting, they are said to be in queue (line). A bank may be used as a familiar analogy to demonstrate this concept. If the customers line up at individual tellers' windows, they form a number of queues, each waiting to be served in turn. One queue may move more rapidly than another, such that customers who arrived later may be served earlier. Alternatively, the bank may employ a single large queue for all customers to be served in turn by the next available teller. In this manner, the customers are served first-in, first-out (FIFO). This latter type of queue is the one normally used in telephone systems.

2.5.1 Queue Length Indication

Frequently, because of traffic engineering's concern with service criteria, only the effects of the queue are considered; that is, average delay or delay distribution. However, it is often necessary to provide an indicator of the waiting load to administrative or maintenance personnel to aid them in determining whether to supplement the servers. If the servers are operators, practices might be put into effect that would defer certain operations and, as a result, reduce the holding time per call rather than add to the operator force. If the servers are equipment, those out of service for routine maintenance could be restored to service, and practices could be put into effect to shorten equipment work time, such as canceling routine on-line diagnostic procedures.

Through the years, many kinds of indicators have been used for these purposes, such as, group busy, timed group busy, single call waiting, and a large number of lamps lit on a multiple switchboard. A better indicator is to keep track of the number of calls in queue and, when this number exceeds a predetermined threshold, provide the indication. The drawback is the fact that the right number to use as an indicator is dependent on the number of servers and, at least in the case where the servers are operators, this number will vary widely during the course of the day. In call-distribution systems (such as those used for airline agents), the system keeps track of the number of operators on duty and adjusts the indicator threshold appropriately.

2.5.2 Queuing Formulas

The assumptions of ordered or random service will have no effect on the average length of a queue or on the probability of a queue of a given length. These assumptions deal only with which waiting call is to be served, not whether or when waiting calls are to be served. However, the constant or exponential holding-time assumptions have a significant bearing on average queue length, and also on the probability of a queue of a given length. Wilkinson has provided us with formulas that predict the average length and distribution of lengths of queues when all servers are busy, and the average distribution of queue lengths during a finite period of time (such as one hour) for the assumptions of infinite sources and exponential holding time (Wilkinson 1953). Crommelin has provided similar formulas for applications with infinite sources and constant holding time, such as intertoll trunk groups (Crommelin 1933).

2.6 PROBABILITY CONCEPTS

The concept of probability is uncertainty—a concept difficult to define, but easy to demonstrate. The result of flipping a coin or cutting a deck of cards involves uncertainty insofar as the outcome is concerned. It is the very inability to confidently predict such outcomes that gives rise to the concept of probability. Feller's two-volume book on probability theory provides a solid introduction to the subject (Feller 1971).

A probability value (P) states the likelihood of occurrence of a specific event; thus, probability distributions are bounded by the values zero and one. This property of probability distributions is used to advantage when it is sometimes easier to compute the probability of an event not occurring. The probability of this alternative is the complement (Q) of P, as defined in (2.6) and (2.7):

$$P = 1 - Q \tag{2.6}$$

$$Q = 1 - P \tag{2.7}$$

where

P = probability an event will occur

Q = probability an event will not occur

The employment of probability calculations has become very important in many areas of science. The use of mathematical statistics and probability to predict the behavior of individual items is accomplished by using observations and measurements made on adequate samples of other items with similar characteristics. The telephone industry has created an entire system designed, operated, and maintained on the strength of probability predictions.

In dealing with large populations, that is, traffic samples in the steady-state condition, it is necessary to deal with probability distributions. These distributions provide the probability associated with every possible event of the particular set of events under consideration, within the limits and constraints of their underlying assumptions.

2.6.1 Binomial Distribution

The binomial distribution considers the probability of sequences in an experiment for which there are only two points in the sample space. It is a discrete distribution (the values of its sample points are discrete, integral numbers) used to compute the probability that a sample of observations (trials) will result in a specified number of successes, given the probability of success for the population. When applied to the telephone industry, the binomial distribution assumes that there are a limited (finite) number of sources and that blocked calls are held up to their normal holding time.

2.6.2 Normal Distribution

As the number of samples increases, the number of possible outcomes increases exponentially, and the binomial distribution approaches the normal distribution as its limit. The normal distribution is completely defined by the mean and standard deviation of the population. When plotted, it results in the familiar bell-shaped curve appropriately known as the normal curve. There is no single normal curve, but a family of curves—all have the same general shape, but may be wider or narrower depending on the standard deviation.

2.6.3 Poisson Distribution

The Poisson distribution is another approximation of the binomial distribution, as indicated by its full name, Poisson's exponential binomial limit (Molina 1942). It is applicable when the number of samples becomes very large (approaches infinity) and the probability goes to zero. The Poisson distribution is completely defined by the value of the mean traffic density, and is the basis for the Poisson formula presented in Section 3.3.1.

REFERENCES

Crommelin, C.D., "Delay Probability Formula," *Journal of Post Office Electrical Engineers*, Vol. 26, 1933, pp. 266–274.

Feller, W., *An Introduction to Probability Theory and its Applications*, 2 Vols., New York: John Wiley and Sons, 1971.

Molina, E.C., *Poisson's Exponential Binomial Limit*, New York: D. van Nostrand, 1942.

Wilkinson, R.I., "Working Curves for Delayed Exponential Calls Served in Random Order," *Bell System Technical Journal*, Vol. 32, No. 2, 1953, pp. 360–383.

Chapter 3
Common-Equipment Dimensioning

Common equipment refers to those devices or facilities (resources) that are used on a shared basis, and are commonly referred to as *servers*. Typical servers employed in a telecommunication system are trunk groups, signaling registers, and operator positions. An adequate number of servers must be available to ensure that requests for service are met within the specified service objective; that is, the probability that a server is unavailable when needed (the grade of service) must be kept within acceptable limits.

From the earliest days of the telephone, mathematicians have concerned themselves with the problem of predicting service under various loading conditions. In 1909, Erlang published his first paper on probability as applied to telephone engineering (Brockmeyer, Halstrom, and Jensen 1948). Many others have expanded and added to his work as new equipment and serving arrangements have been devised.

A number of traffic formulas and tables have been derived that adequately fit, or can be adapted to fit, most practical situations within the telephone industry. The major traffic formulas, based on the Poisson, Erlang-B, and Erlang-C distributions, plus those based on the Engset distribution, are included in this book. The emphasis is on an understanding of how to properly select and use the various formulas.

Computer programs written in BASIC are provided for these traffic formulas. These programs can be very useful for interpolating between table values that are not representative of the system being analyzed, or for determining more precise values for a specific application. They are intended to be used with typical personal computers, and employ a user-friendly, interactive format. Prompts and responses generated by the computer in the examples given are underlined for clarification.

3.1 APPLICABLE ASSUMPTIONS

A traffic system consists of a set of time-varying demands, a group of facilities to serve those demands, and service criteria to be met by the system. A mathematical model consists of a set of assumptions about the first two factors, with the objective of predicting the third. It is important to remember that traffic theory is based upon probability theory and assumptions. Assumptions should be minimized, but those based on experience, valid statistical data, and careful analysis can yield satisfactory results. Even using the most sophisticated computer to process the most extensive data, the solution to a traffic problem is still nothing more than a prediction of the probability of blocking or delay.

When traffic formulas, tables, or programs are used, we must understand the underlying assumptions of what distinguishes one from another. The alternative assumptions regarding finite or infinite sources, and the alternatives regarding the disposition of blocked calls, lead to different formulas, as summarized in Table 3.1. In addition, these traffic formulas assume that subscribers originate calls at random, and independently of other subscribers.

The failure of an assumption to be applicable to the system under consideration can result in a significant error in the engineering of the system. Furthermore, the validity of any assumption with respect to the particular system being engineered must be determined, not by the mathematician who derived the formula, but by the person who intends to apply it. Misapplication of traffic formulas is often a result of an unfamiliarity with the underlying assumptions. Such misapplication may lead to using a table intended for one purpose for some other purpose, or using it for its intended purpose in a particular application where the actual conditions are not those that were considered in the table development.

In many cases the grade of service level is so low that the offered traffic loads permitted by each of the major traffic formulas differ negligibly, and use of the wrong formula would not appreciably affect the result. If, however, engineering is done at a higher level of blocking, the Erlang-B distribution differs significantly from the Poisson and the Erlang-C distributions. The difference in grade of service as predicted by the three formulas reflects the fact that under the Poisson and Erlang-C assumptions blocked calls can contribute to the congestion of the group. Under the Erlang-B blocked-calls-cleared assumption, this is impossible.

3.1.1 Independent Sources

The subscribers (sources of calls) are independent as far as originating calls are concerned. This assumption seems fairly reasonable, because it

Table 3.1 Summary of Traffic Formula Assumptions

Traffic Formula	Number of Sources	Blocked-Call Disposition	Holding-Time Distribution
Poisson	Infinite	Held	Constant or Exponential
Erlang-B	Infinite	Cleared	Constant or Exponential
Erlang-C	Infinite	Delayed	Exponential
Crommelin	Infinite	Delayed	Constant
Binomial	Finite	Held	Constant or Exponential
Engset	Finite	Cleared	Constant or Exponential
Delay	Finite	Delayed	Exponential

is difficult to imagine how one subscriber can know how many others are using their telephones, or how the subscriber's actions can be regulated thereby. The fact that calls may be made from subscribers in a group to others in the same group (communities of interest), and the fact that subscribers cannot originate calls while they are busy with incoming calls, do make them somewhat dependent on the actions of others. However, the effect of this is relatively small and may be neglected.

3.1.2 Random Calls

The term "at random" has been established by general agreement; that is, by definition. The origination of calls by customers is due to definite causes, and normally are independent of calls initiated by other subscribers. However, during periods of flood, panic, disaster, and the like, the traffic does not originate at random. It comes in bunches (peaked) because a single cause has impelled a large number of people to make calls nearly simultaneously. In such cases it is not appropriate to use the random assumption. An assumption for this type of traffic would require a larger number of trunks or circuits for a given service criterion than would the random assumption.

3.1.3 Server-Holding Times

Many studies show that a good curve fit to the variation in holding times can be made by use of the negative power of the exponential (see Section 2.1.2). For estimating the probabilities of blocking or delay congestion, the result of substituting a constant holding time equal to the average of varying holding times seems to be of negligible concern from a theoretical standpoint. Studies made on actual calls also indicate that these effects are of a secondary order when considering only the question of how many calls are lost or delayed. However, in predicting the probability of the duration of the delays, the effects of holding-time variations are significant and cannot be neglected.

3.1.4 Immediate Connection

In general, the time required to set up a connection is small when compared with the associated call-holding time; therefore, it may often be ignored. When dealing with delay-service criteria that include connection time, this time would have to be added to the computed delays. In addition, when short-duration data messages are considered, the time required to set up a connection may be an important factor that must be taken into account.

3.1.5 Number of Sources

The number of sources that can make a demand on a particular group of trunks, circuits, or other resources (servers) has a definite bearing on the service these sources can expect to obtain. For example, consider a system consisting of only one source and one server. The source could never make a new demand at a time when the server was busy. As a result, the probability of being blocked or delayed is zero (nonblocking service). If another source is added, either of the sources will be blocked or delayed if the other happens to be using the server, and the probability of congestion is no longer zero. Carried further, as the number of sources is increased, keeping the total load constant, the probability of congestion continues to increase. The effect of adding sources diminishes rapidly, however, and a point is soon reached where there is negligible difference in the probability of congestion regardless of how many sources are added.

Mathematicians have taken advantage of this phenomenon in the major traffic formulas, which assume infinite sources (worst case for blocking). In most applications the number of sources in relation to the number of servers is very large; therefore, the assumption is valid. It simplifies the mathematical formulas and minimizes the number of tables required. However, there are cases (such as line concentrators) where the limited source effect may be significant and cannot be ignored.

3.1.6 Disposition of Blocked Calls

Servers are seldom provided so liberally that every call can always find an idle path. The question then arises, what happens to calls that find all servers busy? This depends on the switching equipment or the operating practices, together with the subscriber's reactions thereto. Many assumptions for the disposition of blocked calls (also referred to as lost calls) can be made, with three common cases following:

a. Blocked Calls Cleared: A call failing to find an idle server is cleared from the system and does not reappear.
b. Blocked Calls Held: If a server is not obtained immediately, the call waits for an interval of time equal to its holding time, and it is then withdrawn; however, if a server becomes idle while the call is waiting, it seizes and holds the server for the interval remaining before its holding time period expires.

c. Blocked Calls Delayed: A call failing to find an idle server waits in queue until a server becomes free, at which time it seizes and holds the server for the full call-holding time (no call is lost under this assumption).

3.1.7 Deviations from Assumptions

Remember that any deviation from these assumptions will cause a difference between the actual and theoretical results. Calls do not always arrive at random. Many things, from a giveaway radio program to a drop in the stock market, cause a nonrandom surge in the calling rate. It is obvious that all holding times are not equal, nor will they fit perfectly on a negative-exponential curve. The assumption that within a system all blocked calls are cleared, held, or delayed can be easily disproved in each case, and sources are never infinite. Yet the above assumptions are sufficient to define mathematical models from which the probability of congestion can satisfactorily be predicted in most cases.

3.2 TRAFFIC CALCULATIONS

Trunk groups and server pools are provided to serve calls on a shared basis. An adequate number of trunks and servers must be provided in each server pool to ensure that the specified service objectives are met. Trunks are assigned to serve calls on an immediate basis, whereas servers may be assigned to serve calls either immediately or on a delayed basis.

3.2.1 Trunk-Group Traffic

Trunks are held from the initiation of a seize sequence until the end of a release sequence (see Section 1.4.6). It does not matter whether information is exchanged over the trunks. Trunk-group traffic, therefore, is the product of the number of calls handled by the group, and the duration of the calls. Equation (3.1) may be used to calculate trunk-group traffic. If the traffic is calculated for the busy hour, then it is defined as the busy-hour traffic expressed in Erlangs.

Trunk-Group Traffic = (Number of Calls)(Call-Holding Time)   (3.1)

For example, consider a trunk group carrying 500 three-minute calls during the normal busy hour. The trunk-group traffic is then

(500 calls/hour)(3 min/call)/(1 hour/60 min) = 25 Erlangs

3.2.2 Signaling-Register Traffic

Signaling registers are held only long enough to provide the required service (during the signaling cycles), and are then returned to the appropriate pool to serve subsequent requests. Traffic levels for various classes of register pools may be derived from empirical data or defined by standard practice. They may also be calculated from specified traffic parameters supplemented by assumptions when the available data are inadequate. Equation (3.2) may be used to calculate register-pool traffic.

$$\text{Register-Pool Traffic} = (A)(T_s)(\text{CAF})/(T_c) \tag{3.2}$$

where

$$A = \text{Total traffic served}$$

$$T_s = \text{Mean register-holding time}$$

$$T_c = \text{Mean call-holding time}$$

$$\text{CAF} = \text{Call-attempt factor}$$

Total traffic served refers to the total offered traffic that requires the services of the specific pool for some portion of the call. For example, a DTMF receiver pool is dimensioned to serve only the DTMF dialing portion of total switch traffic generated by sources that use DTMF signaling. The call-holding time normally includes the signaling time in this equation. Otherwise, the register-holding time must be added to the denominator.

Consider a private switchboard (PABX) where subscribers dial other extensions using a 4-digit number. If DTMF dialing is used, and receiver-holding time is derived using Table 1.3, the typical DTMF receiver-holding time is found to be 5.2 seconds. In the event that a register pool serves traffic with differing signaling parameters, the mean register-holding time may be calculated using (3.3):

$$\text{Mean Register-Holding Time} = a(T_1) + b(T_2) + \ldots + k(T_n) \tag{3.3}$$

where

$$a, b, \ldots, k = \text{Fractions of total traffic served}$$

$$T_1, T_2, \ldots, T_n = \text{Individual register-holding times}$$

Consider a CO where subscribers dial local calls using a 7-digit (NNX-XXXX) number, and toll calls using an 11-digit (1-NYX-NNX-XXXX) number. If 10 percent of the calls are toll calls, and the remainder are to local subscribers, DTMF dialing is used, and receiver-holding times are derived using Table 1.3, the mean DTMF receiver-holding time may be calculated as follows:

Mean Receiver-Holding Time = (12.0 s)(0.1) + (8.1 s)(0.9) = 8.49 s

3.3   POISSON DISTRIBUTION

The Poisson distribution assumes infinite sources, blocked calls are held, and is applicable for either constant or exponential holding time. It is used in North America for dimensioning PABX-to-CO and other final trunk groups (blocked calls not offered to an alternate route). We might well question the use of the Poisson distribution with its blocked calls held assumption for trunking when, in fact, a call arriving in the system and not finding an idle trunk is usually given an ATB (all trunks busy) signal, and is not permitted to wait a holding time or any other period of time. The justification for its usage lies in the fact that studies have shown that the Poisson distribution produces a blocking probability that closely matches the actual results for final trunk groups.

3.3.1   Poisson Formula

The Poisson formula involves a summation from the number of trunks in the trunk group to infinity, which is not amenable to easy calculation. Taking advantage of the fact that the sum of all terms in a probability distribution must equal one (see Section 2.6), it is far easier to calculate the summation from zero to one less than the number of trunks, then subtract that value from one, as shown in (3.4):

$$P = \sum_{i=N}^{\infty} \frac{A^i e^{-A}}{i!} = 1 - \sum_{i=0}^{N-1} \frac{A^i e^{-A}}{i!} \tag{3.4}$$

where

N = Number of servers in group
A = Traffic offered to group
e = Natural logarithm base (≈ 2.7183)

The following program may be used to compute (3.4):

```
100  REM USE FOR POISSON FORMULA
110  INPUT "ENTER NUMBER OF SERVERS (N)";N
120  INPUT "ENTER OFFERED TRAFFIC (A)";A
130  LET X=1
140  FOR I=1 TO (N-1)
150  LET Y=1
160  FOR J=1 TO I
170  LET Y=A/J*Y
180  NEXT J
190  LET X=X+Y
200  NEXT I
210  LET Q=X*EXP(-A)
220  PRINT USING "P=#.####";1-Q
230  END
```

For example, use the program to compute the blocking probability for a 24-channel final trunk group with offered traffic of 15 Erlangs:

```
RUN POISSON
ENTER NUMBER OF SERVERS (N)
? 24
ENTER OFFERED TRAFFIC (A)
? 15
P=0.0195
```

3.3.2 Poisson Traffic Capacity Tables

Appendix A contains traffic capacity tables based on the Poisson formula. To use these tables for solving the above problem, select the row where $N = 24$ and read over to the right until 15 or more Erlangs is found. Then read the blocking probability at the top of that column—in this case 0.02 for 15.1 Erlangs. If the problem were to determine how many trunks would be needed to handle 15 Erlangs of traffic at P.02, then start at the 0.02 column and read down until 15 or more Erlangs is found. Then read over to the end of that row—in this case the row for $N = 24$.

3.4 ERLANG-B DISTRIBUTION

The Erlang-B distribution is based on the assumptions that sources are infinite, blocked calls are cleared, and holding time is either constant or exponential. It is used to dimension server pools that are not provided with a waiting queue (immediate service), such as interswitch trunk groups. In North America it is used mainly for dimensioning high-usage trunk groups in alternate-routing arrangements (block calls offered to another high-usage group or a final group).

3.4.1 Erlang-B Formula

The Erlang-B formula, also called the *Erlang formula of the first kind*, is given in (3.5):

$$P = E_1(N, A) = \frac{\dfrac{A^N}{N!}}{\displaystyle\sum_{i=0}^{N}\frac{A^i}{i!}} \tag{3.5}$$

where

N = Number of servers in group

A = Traffic offered to group

The following program may be used to compute (3.5):

```
100  REM USE FOR ERLANG-B FORMULA
110  INPUT "ENTER NUMBER OF SERVERS (N)";N
120  INPUT "ENTER OFFERED TRAFFIC (A)";A
130  LET X=1
140  FOR I=1 TO N
150  LET X=A/I*X
160  NEXT I
170  LET Y=1
180  FOR J=1 TO N
190  LET Z=1
200  FOR K=1 TO J
```

```
210  LET Z=A/K*Z
220  NEXT K
230  LET Y=Y+Z
240  NEXT J
250  LET P=X/Y
260  PRINT USING "P=#.####";P
270  END
```

For example, use the program to compute the blocking probability for a 24-channel high-usage trunk group with offered traffic of 15 Erlangs:

```
RUN ERLANG-B
ENTER NUMBER OF SERVERS (N)
? 24
ENTER OFFERED TRAFFIC (A)
? 15
P=0.0084
```

3.4.2 Erlang-B Traffic Capacity Tables

Appendix B contains traffic capacity tables based on the Erlang-B formula. To use these tables for solving the above problem, select the row where $N = 24$ and read over to the right until 15 or more Erlangs is found. Then read the blocking probability at the top of that column—in this case 0.01 for 15.3 Erlangs. If the problem were to determine how many servers would be needed to handle 15 Erlangs of traffic at P.01 (also called B.01), then start at the 0.01 column and read down until 15 or more Erlangs are found. Then read over to the end of that row—in this case the row for $N = 24$.

Telephone operating companies use alternate routing tables based on the Erlang-B formula for dimensioning high-usage trunk groups engineered for significant overflow. They also use peakedness tables (also known as Wilkinson or Neal-Wilkinson tables) based on the Erlang-B formula for dimensioning trunk groups with peaked traffic. Peakedness in an offered load can be caused by alternate routing or by other system variances such as higher traffic when the rates are lower. The peakedness factor (Z) is the traffic variance-to-mean ratio—when $Z < 1$ traffic is said to be smooth, when $Z = 1$ traffic is said to be random, and when $Z > 1$ traffic is said to be peaked.

3.5 ERLANG-C DISTRIBUTION

The Erlang-C distribution assumes sources are infinite, blocked calls are delayed, and holding times are approximated by a negative-exponential distribution. It is typically used for dimensioning common-equipment server pools in which call attempts wait in a FIFO buffer (queue) until an idle server is available.

3.5.1 Erlang-C Formulas

The Erlang-C formula given in (3.6), also called the *Erlang formula of the second kind,* defines the probability that a call will be delayed for a time greater than zero. Equation (3.7) defines the probability that a call will be delayed longer than a given time. Equations (3.8) and (3.9) may be used to compute the average delay on all calls and the average delay on calls delayed, respectively.

$$P(>0) = E_2(N,A) = \frac{\dfrac{A^N N}{N! \, (N-1)}}{\displaystyle\sum_{i=0}^{N} \dfrac{A^i}{i!} + \dfrac{A^N N}{N! \, (N-1)}} \tag{3.6}$$

$$P(>t) = P(>0)e^{-(N-A)T_1/T_2} \tag{3.7}$$

$$D_1 = P(>0)T_2/(N-A) \tag{3.8}$$

$$D_2 = T_2/(N-A) \tag{3.9}$$

where

N = Number of servers in group

A = Total offered traffic

T_1 = Acceptable delay time

T_2 = Mean server-holding time

e = Natural logarithm base (≈ 2.7183)

The following program may be used to compute (3.6) through (3.9), in-clusive:

```
100  REM USE FOR ERLANG-C FORMULA
110  INPUT "ENTER NUMBER OF SERVERS (N)";N
120  INPUT "ENTER OFFERED TRAFFIC (A)";A
130  INPUT "ENTER ACCEPTABLE DELAY TIME (T1)";T1
140  INPUT "ENTER SERVER-HOLDING TIME (T2)";T2
150  LET X=1
160  LET Y=0
170  FOR I=1 TO N
180  LET Y=X+Y
190  LET X=A/I*X
200  NEXT I
210  LET P0=((N/(N-A))*X)/(Y+((N/(N-A))*X))
220  LET P1=P0*EXP(-(N-A)*T1/T2)
230  LET D1=P0*T2/(N-A)
240  LET D2=T2/(N-A)
250  PRINT USING "P(>0)=#.####";P0
260  PRINT USING "P(>T)=#.####";P1
270  PRINT USING "D1=#.####";D1
280  PRINT USING "D2=#.####";D2
290  END
```

For example, use the program to compute the probability of delay greater than zero, the probability of delay greater than 3 seconds, the average delay on all calls, and the average delay on calls delayed for a 24-server common-equipment pool with offered traffic of 20 Erlangs, assuming mean server-holding time of 6 seconds:

```
RUN ERLANG-C
ENTER NUMBER OF SERVERS (N)
? 24
ENTER OFFERED TRAFFIC (A)
? 20
ENTER ACCEPTABLE DELAY TIME (T1)
? 3
```

```
ENTER SERVER-HOLDING TIME (T2)
? 6
P(>0)=0.2981
P(>T)=0.0403
D1=0.4471
D2=1.5
```

3.5.2 Erlang-C Delay Loss Probability Tables

Appendix C contains delay loss probability tables based on the Erlang-C formulas. To use these tables for solving the above problem, start with the table for $N = 24$ and select the row where $A = 20$ and read the value of $P(>0)$ in the first column (0.2981). Then read over to the column for $T_1/T_2 = 0.5$ (3 sec./6 sec.) and read the value of $P(>t)$ which is 0.0403. The values of D_1 and D_2 may be calculated using (3.8) and (3.9). If the problem were to determine the level of traffic that could be handled by the server group with a delay loss probability less than 0.05, then read down the 0.5 column until 0.0480 is found, and over to the end of that row—in this case the row for $A = 20.2$.

3.6 ENGSET DISTRIBUTION

The Engset distribution is based on the assumptions that sources are finite, blocked calls are cleared, and holding time is either constant or exponential. It is typically used for dimensioning equipment that has a limited number of sources, such as line concentrators.

3.6.1 Engset Formula

The Engset formula, given in (3.10), is indicative of the increased mathematical complexity encountered when the number of sources are limited (finite):

$$P = \cfrac{\dfrac{(S-1)!}{N!\,(S-1-N)!}\left[\dfrac{A}{S-A(1-P)}\right]^N}{\displaystyle\sum_{i=0}^{N}\dfrac{(S-1)!}{i!\,(S-1-i)!}\left[\dfrac{A}{S-A(1-P)}\right]^i} \qquad (3.10)$$

where

S = Number of sources

N = Number of servers in group

A = Traffic offered to group

Note that the Engset formula includes P on both sides of the equation, which requires an iterative (trial and error) process to obtain a solution. Equation (3.11) can be used to obtain a good approximation if P is very small (as it usually is). That is, as P decreases, the value of one minus P approaches unity $(1 - P \approx 1)$, and $S - A(1 - P)$ approaches $S - A$.

$$P \approx \frac{\dfrac{(S-1)!}{N!\,(S-1-N)!}\left[\dfrac{A}{S-A}\right]^N}{\displaystyle\sum_{i=0}^{N}\dfrac{(S-1)!}{i!\,(S-1-i)!}\left[\dfrac{A}{S-A}\right]^i} \qquad (3.11)$$

The following program may be used to compute (3.11):

```
100   REM USE FOR APPROXIMATE ENGSET FORMULA
110   INPUT "ENTER NUMBER OF SOURCES (S)";S
120   INPUT "ENTER NUMBER OF SERVERS (N)";N
130   INPUT "ENTER OFFERED TRAFFIC (A)";A
140   LET X1=1
150   FOR J1=(S-N) TO (S-1)
160   LET X1=X1*J1
170   NEXT J1
180   LET X2=1
190   FOR J2=1 TO N
200   LET X2=(X2*A/(S-A))/J2
210   NEXT J2
220   LET X=X1*X2
230   LET Y=1
240   FOR I=1 TO N
250   LET Y1=1
```

```
260  FOR K1 = (S − I) TO (S − 1)
270  LET Y1 = Y1*K1
280  NEXT K1
290  LET Y2 = 1
300  FOR K2 = 1 TO I
310  LET Y2 = (Y2*A/(S − A))/K2
320  NEXT K2
330  LET Y = Y + Y1*Y2
340  NEXT I
350  LET P = X/Y
360  PRINT USING "P = #.####";P
370  END
```

For example, use the program to compute the approximate loss probability for a line concentrator with 60 subscribers (sources), 24 channels (servers), and offered traffic of 15 Erlangs (0.25 Erlangs per subscriber):

```
RUN ENGSET
ENTER NUMBER OF SOURCES (S)
? 60
ENTER NUMBER OF SERVERS (N)
? 24
ENTER OFFERED TRAFFIC (A)
?15
P = 0.0032
```

3.6.2 Engset Loss Probability Tables

Appendix D contains loss probability tables based on the Engset formula. To use these tables for solving the above problem, start with the table for $S = 60$, select the row where $A = 15$, read over to the column for $N = 24$, and read the value of $P = 0.0032$. If the problem were to determine the level of traffic that could be handled by the line concentrator with a loss probability less than 0.005, then read down the 24 column until 0.0032 is found, and over to the end of that row—in this case the row for $A = 15$. Note that the loss probability for 15.6 Erlangs of 0.0052 is close, but greater than 0.005. An offered load of about 15.5 Erlangs would lead to a loss probability of about 0.005 for this problem.

REFERENCE

Brockmeyer, E., H. Halstrom, and A. Jensen, "The Life and Times of A.K. Erlang," *Transactions of the Danish Academy of Sciences,* No. 2, Copenhagen, 1948.

Chapter 4
Switching Matrix Congestion

Grade of service (blocking probability) in switching matrices is generally called congestion, and congestion standards have been defined for various classes of switches. *Strictly nonblocking matrices* introduce no blocking at any traffic load. That is, they have zero congestion at all occupancy levels up to and including one Erlang per channel. A matrix may be defined as nonblocking in the wide sense if all possible blocking states can be avoided by the judicious selection of paths (links) for new calls (Benes 1962).

Essentially nonblocking matrices are those that exhibit congestion levels that are negligible at a given traffic level or for a specific application. A *quasi-nonblocking matrix* is defined as one with congestion much less than 0.001 at the anticipated busy-hour load (Duerdoth and Seymour 1973). A *noncongestive matrix* is defined as one with congestion less than 0.01 when the inlets have a binomial occupancy of 0.98 Erlangs, and less than 0.001 when the inlets have a binomial occupancy of 0.95 Erlangs (Karnaugh 1976). A *virtually nonblocking matrix* is defined as one with a busy-hour congestion level much less than that required for the particular application (Boucher 1982). This differs from the quasi-nonblocking and noncongestive definitions given above in that it is wholly dependent on the parameters specified for the application (it does not rely on arbitrarily specified congestion or traffic levels).

4.1 SWITCHING MATRIX CONCEPTS

Switching matrices are used to interconnect inlet and outlet terminations to each other and to service circuits. They may be arranged in different ways to provide for connections between the same number of terminations, or to provide for additional terminations. Switching matrices are often referred to as switching networks; however, in this book the term *network* is used to refer to systems such as the PSTN, ISDN, and LANs, whereas the terms *matrix* and *matrices* are used to refer to the switch arrays within the switching equipment.

There are four factors that weigh heavily upon the design of switching matrices, and cause a wide variety of systems to evolve. These are the number of inlets and outlets, the blocking objectives and traffic characteristics, the cost of manufacturing and packaging the switch arrays, and the resulting cost of control. Often, the differences in congestion characteristics and cost are significant, and the design of multistage matrices has received considerable study. Normally, a design center will be selected that represents the typical parameters to be imposed upon the system. Given these parameters, it is possible to determine the optimum design (least cost to meet the requirements).

Computer programs written in BASIC are provided for calculating congestion in typical switching matrices. They are intended for use with typical personal computers, and employ a user-friendly, interactive format. Prompts and responses generated by the computer in the examples given are underlined for clarification.

4.1.1 Space-Divided Switch Array

The basic building block of a space-divided switching matrix is the switch array, as shown in Figure 4.1 with its typical block diagram. Such an array provides for the interconnection of any of N inlets to any of M outlets (full availability). In this case, a connection is shown from inlet 2 to outlet 3 and from inlet 3 to outlet 2 to complete a two-way connection. The switch hardware may be a crossbar, reed relay, or solid-state switch. There are a considerable number of these devices, with varying degrees of reliability, compactness, cost, and speed of operation. However, the specific hardware is not of importance in this discussion—only matrix concepts are—and the switch contacts are referred to as *crosspoints*, regardless of the type provided.

4.1.2 One-Stage Matrix

There are always many possible matrix configurations for a set of switching parameters. For example, an N-by-N (N × N) array, as depicted in Figure 4.2, might be designed to provide a one-stage switching matrix. This is a nonblocking configuration, but is relatively inefficient because only N of the crosspoints could ever be used at one time.

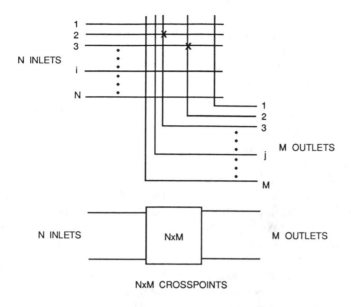

NxM CROSSPOINTS

(WHERE X = CROSSPOINT CONNECTION)

Figure 4.1 Typical Switching Array

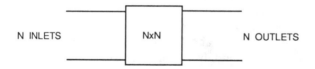

Figure 4.2 One-Stage Switching Matrix Diagram

4.2 LINK SYSTEMS

Link systems employing two or more matrix stages are more common for moderate to large switching systems. Matrices with up to four stages are described in this book. The same principles and analysis techniques may be used with larger matrices.

When analyzing switching matrix congestion (internal blocking) in space-divided link systems, it is first necessary to define the conditions under which blocking can occur, and then to formulate a technique whereby the degree of congestion can be determined. The two conditions that cause congestion in a link system are a) applicable outlets are all busy, and b) available idle links cannot be matched. The first condition can occur if there is a concentration or expansion in the array. The second condition, called *idle-link mismatch* or *matching loss,* can occur only if there are at least three stages of switching.

The most convenient tool for analyzing link systems is the linear graph method developed by Lee (Lee 1955). A linear graph, also known as a channel graph, is merely a simplified diagram of the links required to establish the connection from an inlet to an outlet. Nodes (circles) represent switching stages, and branches (lines) represent the links that interconnect the stages.

Traffic density, expressed in Erlangs, is equal to the average number of busy links. For example, if twenty randomly accessed links carry a load of 15 Erlangs, two important facts are known about the group—the average number of busy links is 15, and the probability that a given link is busy is 0.75. The latter is derived because, on the average, 15 of the 20 links (75%) are busy at any one time; therefore, the probability of any link being busy at any given instant is 0.75.

The assumptions made in developing the following congestion equations are binomial distribution of occupied links in each stage, independent link states on each stage and each parallel path, and uniform distribution of traffic on links within each stage.

4.2.1 Two-Stage Link Matrix

The two-stage link matrix depicted in Figure 4.3 has M inlet arrays each containing N inlets, and M outlet arrays each containing N outlets. These arrays are commonly called inlet and outlet groups. This matrix topology requires fewer crosspoints for the same number of terminations as the one-stage matrix, and is also used to provide for more terminations than could reasonably be accommodated by a one-stage matrix.

The linear graph is simply two nodes with one branch connecting them, because there is only one possible path between any A-stage inlet and B-stage outlet. If P is the probability that a call from an inlet to an idle outlet will be blocked, and Q is its complement (probability that the call will not be blocked), then

$$P = P(\text{A-B link is busy})$$

$$P = \text{Occupancy on the A-B link}$$

$$P = \text{A-B link traffic density in Erlangs}$$

$$P = a \tag{4.1}$$

$Q = P(\text{A-B link is idle})$

$Q = 1 - P = 1 - a$ (4.2)

where

a = Link traffic density = $(A)(N)/M$ Erlangs

A = Mean offered traffic per inlet

N = Number of inlets per group

M = Number of interstage links

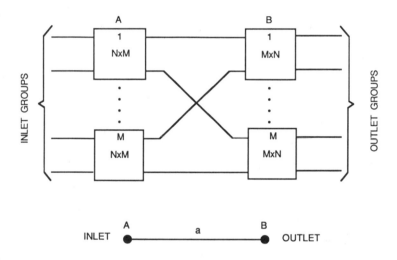

Figure 4.3 Two-Stage Switching Matrix Diagram

If the level of congestion is unacceptable, larger arrays may be used to construct a switching matrix such as that of Figure 4.4. In this case, two parallel paths (interstage links) are available between any A-stage inlet and B-stage outlet, and each path will carry half of the traffic (assumption of uniform traffic distribution). The linear graph, then, has two branches between the nodes, and both links must be busy before a call will be blocked. Using the probability for independent events occurring simultaneously, the following analysis may be used to predict the probability of congestion for this matrix.

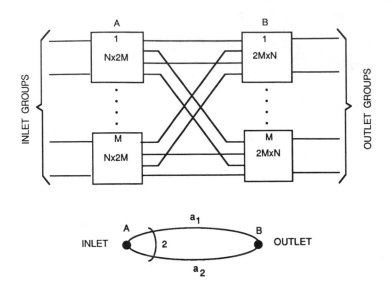

Figure 4.4 Two-Stage Switching Matrix with Multiple Links

$$P = P(\text{link } a_1 \text{ and } a_2 \text{ busy})$$

$$P = P(\text{link } a_1 \text{ busy})P(\text{link } a_2 \text{ busy})$$

$$P = (a_1)(a_2) = a^2 \tag{4.3}$$

$$Q = P(a_1 \text{ or } a_2 \text{ idle})$$

$$Q = 1 - P = 1 - a^2 \tag{4.4}$$

where

$$a = \text{Link traffic density} = (A)(N)/(2)(M)$$

$$A = \text{Mean offered traffic per inlet}$$

$$N = \text{Number of inlets per group}$$

$$M = \text{Number of interstage links}$$

Equations (4.5) and (4.6) are general expressions for congestion in a two-stage, multiple-link matrix:

$$P = a^X \tag{4.5}$$

$$Q = 1 - a^X \tag{4.6}$$

where

a = Link traffic density = $(A)(N)/(M)(X)$

A = Mean offered traffic per inlet

N = Number of inlets per group

M = Number of interstage links

X = Multiple-link factor

4.2.2 Three-Stage Link Matrix

The three-stage matrix depicted in Figure 4.5 has G inlet groups each containing N inlets, and G outlet groups each containing N outlets. It requires more crosspoints for the same number of terminations as the two-stage matrix; however, the call completion capability of the matrix is considerably improved. The linear graph for this matrix shows M interstage links between any A-stage inlet and B-stage outlet. There will be blocking in the matrix whenever idle A-B links cannot be matched with idle B-C links; that is, if either one of the two links in each path is busy, a call will be blocked (matching loss).

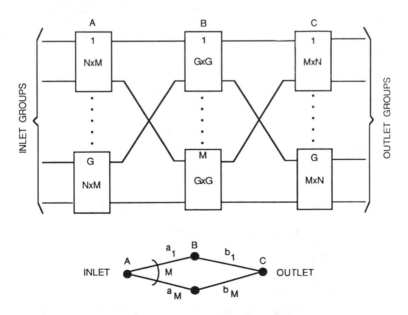

Figure 4.5 Three-Stage Switching Matrix Diagram

Using the probability of several alternate independent events that are not mutually exclusive occurring, the following analysis may be used to solve for the probability of congestion in this matrix. Consider just one path, say path i, involving A-B link a_i and B-C link b_i. The blocking probability for any one of the M paths may be represented as P_i.

$$P = P(a_1 \text{ or } b_1 \text{ busy}) \, P(a_2 \text{ or } b_2 \text{ busy}) \ldots P(a_M \text{ or } b_M \text{ busy})$$

$$P_i = P(a_i \text{ or } b_i \text{ busy}) = P(a_i \text{ busy}) + P(a_i \text{ idle}) \, P(b_i \text{ busy})$$

$$= a_i + (1 - a_i)b_i$$

where

$$(1 - a_i) = P(i\text{th A-B link is idle})$$

If $a_1 = a_2 = \ldots = a_M = a_i$, and $b_1 = b_2 = \ldots = b_M = b_i$, then,

$$P = (P_1)(P_2) \ldots (P_M) = P_i^M = [a + (1 - a)b]^M$$

The common form of the expression for congestion in a three-stage matrix is given in (4.7), and can be read simply as "the probability of blocking is equal to one minus the probability that both links in all possible paths will not be busy simultaneously." Clos has shown that a three-stage link matrix will be strictly nonblocking if the number of links (M) is at least twice the number of inlet terminations minus one ($M \geqslant 2N - 1$) (Clos 1953). Equation (4.7) should not be used for such a matrix, because it will predict a finite, albeit very small, blocking probability.

$$P = [1 - (1 - a)(1 - b)]^M \tag{4.7}$$

for $M < 2N - 1$, where

a = A-B link traffic density in Erlangs = $(A)(N)/M$ Erlangs

b = B-C link traffic density in Erlangs = $(A)(N)/M$ Erlangs

A = Mean offered traffic per inlet

N = Number of inlets per group

M = Number of interstage links

The above example assumed only one outlet could be used to complete a call. In many cases this is true, as in completing a call to a particular subscriber's line. However, if the call is to a trunk group, any idle trunk in the group will serve, and retrials may be initiated. This is simply the selection of another outlet and reattempting to make a connection. In most cases the inlet will look at the same set of A-B links but will now be able to attempt a match with a different set of B-C links as illustrated in Figure 4.6. Because the two outlets are essentially the same (trunks in the same trunk group or lines in the same line-hunting group), the B-C links have been effectively doubled.

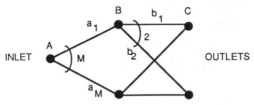

Figure 4.6 Graph for Three-Stage Matrix with Retrial

If P is the probability that a call from an inlet to two idle outlets will be blocked, and P' is the probability that a call from an A-B link to two idle outlets will be blocked, then,

$$P' = P(b_1 \text{ busy})P(b_2 \text{ busy}) = (b_1)(b_2) = b^2$$
$$P = [a + (1 - a)P']^M = [a + (1 - a)b^2]^M$$

By making one retrial, in the event that the first trial is blocked, the grade of service has been improved without adding more crosspoints. Retrials are normally employed by common-control machines, and they impose additional holding time on the common equipment. Equation (4.8) is the expression for a three-stage matrix with multiple trials:

$$P = [a + (1 - a)b^T]^M = [1 - (1 - a)(1 - b^T)]^M \qquad (4.8)$$

where

a = A-B link traffic density = $(A)(N)/M$ Erlangs

b = B-C link traffic density = $(A)(N)/(M)(T)$ Erlangs

A = Mean offered traffic per inlet

N = Number of inlets per group

M = Number of interstage links

T = Number of trials

Note that when $T = 1$ Equation (4.8) reduces to Equation (4.7); therefore, (4.8) is a general expression for congestion in a three-stage matrix. The following program can be used to compute (4.7) and (4.8):

```
100   REM USE FOR 3-STAGE MATRIX
110   INPUT "ENTER NUMBER OF INLETS/GROUP (N)";N
120   INPUT "ENTER NUMBER OF LINKS (M)";M
130   INPUT "ENTER OFFERED TRAFFIC (A)";A
140   INPUT "ENTER NUMBER OF TRIALS (T)";T
150   IF M<2*N-1 GOTO 180
160   PRINT "P=0 (NONBLOCKING)"
170   GOTO END
180   LET A=A*N/M
190   LET B=A/T
200   LET P=(1-(1-A)*(1-B^T))^M
210   PRINT USING "P=##·##^^^^";P
220   END
```

For example, use the program to determine congestion for a three-stage link matrix with 12 inlets per group, 16 interstage links, 0.7 Erlangs per inlet, and one retrial is attempted:

```
RUN 3-STAGE
ENTER NUMBER OF INLETS/GROUP (N)
? 12
ENTER NUMBER OF LINKS (M)
? 16
ENTER OFFERED TRAFFIC/INLET (A)
? .7
ENTER NUMBER OF TRIALS (T)
? 2
P = 8.77E-05
```

where $8.77E-05 = 0.0000877$.

4.2.3 Four-Stage Mesh Matrix

Four-stage mesh matrices, as shown in Figure 4.7, are used for larger switches where a number of frames (switching modules) must be interconnected to provide for more terminations. More crosspoints are required, and now the control unit must select a path through the matrix. This means increased cost, but considerable advantages are gained—they serve more terminations, carry more traffic, and eliminate the need to consider effective availability at the outlets. One characteristic should be noted—mesh matrices have a distinct dependence upon particular links in successive stages; that is, a particular A-B link is associated with a particular C-D link by one particular B-C link. Again, idle-link mismatch can occur across the interstage links. In this case there are M interstage (B-C) links, each of which may be analyzed as for three-stage link matrices (see Section 4.2.2) to yield Equation (4.9):

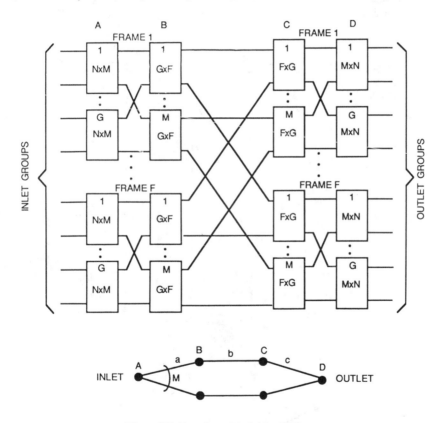

Figure 4.7 Four-Stage Mesh Matrix Diagram

$$P = [1 - (1 - a)(1 - b)(1 - c)]^M \qquad (4.9)$$

where

a = A-B link traffic density = $(A)(N)/M$ Erlangs

b = B-C link traffic density = $(A)(N)/M$ Erlangs

c = C-D link traffic density = $(A)(N)/M$ Erlangs

A = Mean offered traffic per inlet

N = Number of inlets per group

M = Number of B-C links

Figure 4.8 depicts a four-stage mesh matrix with multiple B-C links. The expression for congestion in this matrix is given in Equation (4.10). Note that when $X = 1$ Equation (4.10) reduces to Equation (4.9).

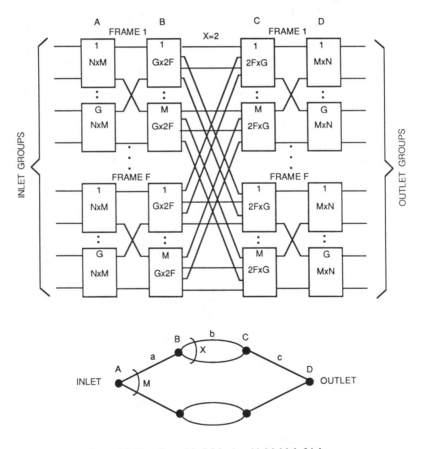

Figure 4.8 Four-Stage Mesh Matrix with Multiple Links

$$P = [1 - (1 - a)(1 - b^x)(1 - c)]^M \qquad (4.10)$$

where

a = A-B link traffic density = $(A)(N)/M$ Erlangs

b = B-C link traffic density = $(A)(N)/(M)(X)$ Erlangs

c = C-D link traffic density = $(A)(N)/M$ Erlangs

A = Mean offered traffic per inlet

N = Number of inlets per group

M = Number of B-C links

X = Multiple-link factor

The effect of retrials can be evaluated in a manner similar to that described for three-stage matrices. Figure 4.9 shows a linear graph for a four-stage mesh matrix with T trials. In this case, the formulation is slightly different. The expression for congestion in this matrix is given in (4.11). It can be shown that when $T = 1$ Equation (4.11) also reduces to Equation (4.9).

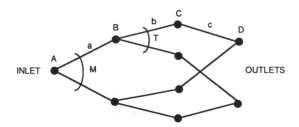

Figure 4.9 Graph for Four-Stage Mesh Matrix with Retrial

$$P = \{1 - (1 - a)[1 - [1 - (1 - b)(1 - c)]^T]\}^M \qquad (4.11)$$

where

a = A-B link traffic density = $(A)(N)/M$ Erlangs

b = B-C link traffic density = $(A)(N)/(M)(T)$ Erlangs

c = C-D link traffic density = $(A)(N)/(M)(T)$ Erlangs

A = Mean offered traffic per inlet

N = Number of inlets per group

M = Number of B-C links

T = Number of trials

Retrials may also be used in combination with multiple links. Consider, for example, the equivalent linear graph of Figure 4.10. The expresssion for congestion in this matrix is given in (4.12):

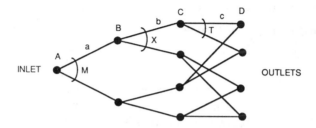

Figure 4.10 Graph for Four-Stage Mesh Matrix With Retrial and Multiple Links

$$P = \{1 - (1 - a)[1 - [1 - (1 - b^X)(1 - c)]^T]\}^M \qquad (4.12)$$

where

a = A-B link traffic density = $(A)(N)/M$ Erlangs

b = B-C link traffic density = $(A)(N)/(M)(X)$ Erlangs

c = C-D link traffic density = $(A)(N)/(M)(T)$ Erlangs

A = Mean offered traffic per inlet

N = Number of inlets per group

M = Number of B-C links

X = Multiple-link factor

T = Number of trials

It can be shown that when $T = 1$ and $X = 1$ Equation (4.12) also reduces to (4.9); therefore, (4.12) is a general expression for congestion in a four-stage mesh matrix. The following program can be used to compute (4.9) through (4.12):

```
100   REM USE FOR 4-STAGE MESH MATRIX
110   INPUT "ENTER NUMBER OF INLETS/GROUP (N)";N
120   INPUT "ENTER NUMBER OF LINKS (M)";M
130   INPUT "ENTER OFFERED TRAFFIC (A)";A
140   INPUT "ENTER MULTIPLE-LINK FACTOR (X)";X
150   INPUT "ENTER NUMBER OF TRIALS (T)";T
160   LET A = A*N/M
170   LET B = A/X
180   LET C = A/T
190   LET P = (1-(1-A)*(1-(1-(1-B^X)*(1-C))^T))^M
200   PRINT USING "P = ##.##^^^^";P
210   END
```

For example, use the program to determine congestion for a four-stage mesh matrix with 12 inlets per group, 16 interstage links, 0.7 Erlangs per inlet, and one retrial is attempted:

```
RUN 4-STAGE
ENTER NUMBER OF INLETS/GROUP (N)
? 12
ENTER NUMBER OF LINKS (M)
? 16
ENTER OFFERED TRAFFIC/INLET (A)
? .7
ENTER MULTIPLE-LINK FACTOR (X)
? 1
ENTER NUMBER OF TRIALS (T)
? 2
P = 5.89E-03
```

where $5.89E-03 = 0.00589$.

4.2.4 Four-Stage Spiderweb Matrix

Further independence of link states can be obtained by a spiderweb design as shown in Figure 4.11. A significant increase in crosspoints is required, but no longer is an A-B link dependent upon a particular C-D

68

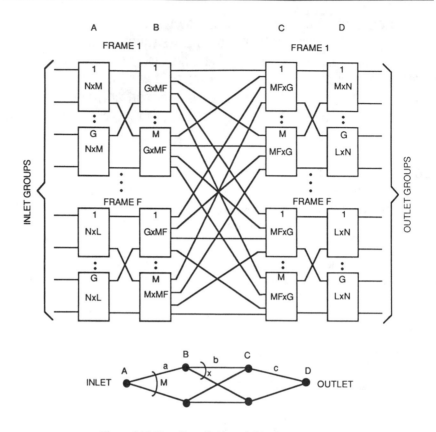

Figure 4.11 Four-Stage Spiderweb Matrix Diagram

link; that is, each A-B link has access to all C-D links. Therefore, the probability of idle-link mismatch is significantly decreased.

In this case, a more complex analytical problem is introduced which does not lend itself to the straight-forward formulation method described for the mesh matrix. The characteristic of previous matrix types was that each intermediate node of the linear graph was connected to only one previous stage node in the graph. Therefore, interstage links could be defined that were independent of each other. The spiderweb matrix, however, increases the number of link combinations by connecting intermediate nodes to more than one previous stage node. The assumption of independent interstage links must then be abandoned.

If P is the probability that a call from an inlet to an idle outlet will be blocked, and P' is the probability that an idle outlet will be blocked from connecting to X idle A-B links, then Figure 4.12 is the subgraph of interest in determining P' (note that the stages have been reversed).

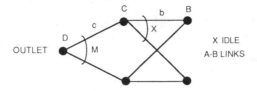

Figure 4.12 Subgraph for Four-Stage Spiderweb Matrix

$$P' = [c + (1 - c)b^X]^M = [1 - (1 - b^X)(1 - c)]^M \qquad (4.13)$$

If Px'' is the probability that there are exactly X idle A-B links then, by definition of the binomial distribution:

$$P''_x = \frac{M!}{X!\,(M - X)!}\, a^{(M-X)}\,(1-a)^X \qquad (4.14)$$

There are $M + 1$ possibilities for the number of idle A-B links, and the overall probability of blocking (P) will be the sum of the alternative events that cause congestion:

$$P = P'_0 P''_0 + P'_1 P''_1 + P'_2 P''_2 + \ldots + P'_M P''_M = \Sigma P'_i P''_i$$

$$P = \sum_{i=0}^{M} \frac{M!}{i!\,(M - i)!}\, a^{(M-i)}(1 - a)^i [1 - (1 - b^i)(1 - c)]^M \qquad (4.15)$$

where

a = A-B link traffic density = $(A)(N)/M$ Erlangs

b = B-C link traffic density = $(A)(N)/M$ Erlangs

c = C-D link traffic density = $(A)(N)/M$ Erlangs

A = Mean offered traffic per inlet

N = Number of inlets per group

M = Number of interstage links

The following program can be used to compute (4.15):

```
100  REM USE FOR 4-STAGE SPIDERWEB MATRIX
110  INPUT "ENTER NUMBER OF INLETS/GROUP (N)";N
```

```
120  INPUT "ENTER NUMBER OF LINKS (M)";M
130  INPUT "ENTER OFFERED TRAFFIC (A)";A
140  LET A=A*N/M
150  LET P=A^M
160  FOR I=1 TO M
170  LET X=1
180  FOR J=(M-I+1) TO M
190  LET X=X*J
200  NEXT J
210  FOR K=1 TO I
220  LET X=X/K
230  NEXT K
240  LET P=P+X*A^(M-I)*(1-A)^I*(1-(1-(A/I)^I)*
     (1-A))^M
250  NEXT I
260  PRINT USING "P = ##.##^^^^";P
270  END
```

For example, use the program to determine congestion for a four-stage spiderweb matrix with 12 inlets per group, 16 interstage links, and 0.7 Erlangs per inlet:

```
RUN SPIDERWEB
ENTER NUMBER OF INLETS/GROUP (N)
? 12
ENTER NUMBER OF LINKS (M)
? 16
ENTER OFFERED TRAFFIC/INLET (A)
? .7
P = 7.49E-05
```

where $7.49E-05 = 0.0000749$.

4.3 TIME-DIVIDED DIGITAL MATRICES

The concept of time-divided digital matrices might seem intimidating at first because suddenly those real, physical connections seem to slip off into abstraction. However, the connection in a time-divided matrix is just as real and as physical as in a space-divided link system. The only significant difference is the speed of the connection, which typically occurs in nano-seconds (ns).

4.3.1 Time-Slot Interchanger

Typical pulse-code modulation (PCM) carrier systems handle 24 channels and generate eight bits of data per time slot. A typical time-division multiplexed (TDM) matrix will group eight PCM carrier systems of 24 channels each (192 total channels). By converting the eight bits per channel PCM carrier time slots from serial to parallel, each of the 192 incoming channels can be handled in 192 time slots. The function of the matrix is to switch (interchange) the time slots; therefore, a one-stage time-divided matrix, as shown in Figure 4.13, is called a *time-slot inter-changer* (TSI).

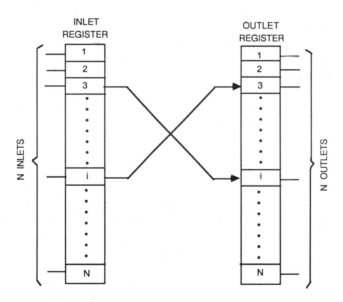

Figure 4.13 Time-Slot Interchanger Diagram

For example, an incoming channel on time-slot 3 may require con-nection to an outgoing channel on time-slot i. In order to accommodate this requirement, an inlet register (buffer) receives the time-slot 3 infor-mation and stores it until time-slot i comes around. At that time, it is transferred to time-slot i of the outlet register for subsequent transmission to the outlet channel. Similarly, information from time-slot i is stored until time-slot 3 comes around again, and is then transferred to provide for two-way traffic.

4.3.2 Multistage Digital Matrices

If no more than 192 channels (circuits) required switching, the above arrangement would be sufficient. If more inlets are introduced, either more time slots or stages are required. The number of time slots are limited due to reproduction quality standards; therefore, a space-divided switching stage is often used to interconnect inlet and outlet TSIs. Such a switching arrangement is called a time-space-time (TST) matrix. Other arrangements with two or more stages (TT, STS, TSTST, *et cetera*) are also used, depending on the requirements of the switching system.

The concept of congestion in multistage time-divided systems is not different than in link systems, but a change in definitions is required. The link systems discussed in Section 4.2 physically connect inlet terminations to outlet terminations. In a multistage time-divided digital matrix, an inlet never passes information directly to an outlet, but rather through interstage registers (buffers). In order for the interstage registers to fulfill their function, they must be able to switch the information in the inlet register time slots to the proper outlet register time slots. It is necessary, then, that the time-slots required by the inlets and outlets be available in the interstage registers at the same time. A call from an inlet is considered lost when no interstage register can be found that has both the inlet and outlet time slots available. Thus, the interstage register time slots are equivalent to interstage links, the resultant block diagram and linear graph are identical to that of the three-stage matrix depicted in Figure 4.5, and (4.7) may be used. However, for the three-stage digital matrix the number of interstage links are equal to the number of interstage time slots. The Clos nonblocking criterion ($M \geqslant 2N - 1$) is also applicable for three-stage digital matrices (see Section 4.2.2).

Digital switches exhibit significantly better grade of service characteristics than equivalent analog switches due to the nature of the time-divided matrix. Rather than assigning matrix links to the connection for the duration of the call, TDM techniques are used to carry multiple calls over each link. Therefore, the effective number of channels is equal to the number of time slots provided for the particular matrix stage. The Jacobaeus equation given below (Jacobaeus 1950), is a more applicable blocking equation for a three-stage (TST or STS) digital matrix. This equation also applies approximately to TSSST, SSTSS, and STSTS matrices provided the central stages (SSS, STS, and TST, respectively) are essentially nonblocking.

$$P = \frac{(N-1)!^2 a^M (2-a)^{(2N-M-2)}}{M! \, (2N-M-2)!} \qquad (4.16)$$

for $N - 1 < M < 2N - 1$
where

N = Number of inlet time slots

M = Number of interstage time slots

a = Mean offered traffic per inlet

Computation of (4.16) typically yields very large and very small numbers which are cumbersome to deal with and may cause overflow or underflow of computer registers. The following recursive form is more amenable to computer manipulation (Boucher 1982):

$$P = a^M (2-a)^{(2N-M-2)} \prod_{i=N}^{M} (i + N - M - 1)/i \qquad (4.17)$$

The following program can be used to compute (4.16) and (4.17):

```
100  REM USE FOR JACOBAEUS EQUATION
110  INPUT "ENTER NUMBER OF INLET TIME SLOTS (N)";N
120  INPUT "ENTER NUMBER OF INTERSTAGE TIME SLOTS (M)";M
130  INPUT "ENTER OFFERED TRAFFIC/INLET (A)";A
140  IF M<2*N-1 GOTO 170
150  PRINT "P = 0 (NONBLOCKING)"
160  GOTO END
170  LET X=1
180  FOR I=N TO M
190  LET X=X*(I+N-M-1)/I
200  NEXT I
210  LET P=X*A^(2*(M-N+1))*(A*(2-A))^(2*N-M-2)
220  PRINT USING "P = ##.##^^^^";P
230  END
```

For example, assume that each switching matrix module for a three-stage digital switch has 192 inlet channels, 192 outlet channels, and 192 interstage time slots. Using the Jacobaeus equation, calculate the grade of service for this matrix if the average busy-hour traffic loading is 0.8 Erlangs per channel:

```
RUN JACOBAEUS
ENTER NUMBER OF INLET TIME SLOTS (N)
? 192
ENTER NUMBER OF INTERSTAGE TIME SLOTS (M)
? 192
ENTER OFFERED TRAFFIC/INLET (A)
? .8
P = 2.73E-04
```

where $2.73E - 04 = 0.000273$.

REFERENCES

Benes, V.E., "Algebraic and Topological Properties of Connecting Networks," *Bell System Technical Journal*, Vol. 36, No. 4, 1962, pp. 1249–1274.

Boucher, J.R., "Is a Nonblocking Digital Matrix Essential, or is Essentially Nonblocking Acceptable?," *IEEE Global Telecommunications Conference*, Vol. 3, Miami, 1982, p. 1309.

Clos, C., "A Study of Nonblocking Switching Networks," *Bell System Technical Journal*, Vol. 32, 1953, pp. 406–424.

Duerdoth, W.T., and C.A. Seymour, "A Quasi-Nonblocking TDM Switch," *International Teletraffic Congress*, Stockholm, 1973, pp. 632/1–4.

Jacobaeus, C., "A Study of Congestion in Link Systems," *Ericcson Technic*, No. 48, Stockholm, 1950, pp. 1–70.

Karnaugh, M., "Design Considerations for a Digital Switch," *International Switching Symposium*, Kyoto, 1976, p. 212-4-2.

Lee, C.Y., "Analysis of Switching Networks," *Bell System Technical Journal*, Vol. 34, 1955, pp. 1287–1315.

Chapter 5
Switching Control Systems

This chapter discusses methods of controlling switching systems and presents functional concepts of the various approaches. The discussion deals exclusively with common-control systems and includes interface components which play major roles in system operation.

5.1 COMMON-CONTROL CONCEPTS

The term *common control* can be used to define a circuit technique and a switching principle. As a technique, it implies use of control which is common to more than one switching device. In order to get the proper perspective for the discussion of various control philosophies and techniques, it is necessary to review the functions the control portion of the system must perform.

On an originating call in a common-control system, the incoming trunk or originating line initiates a call for service and is connected to a receiving device, and a dial tone or start-dial signal is transmitted. When information has been received, a translation takes place indicating the desired disposition of the call. Assuming the called line or trunk route is available, the matrix control device (marker) operates the switching-matrix crosspoints to connect the appropriate inlet and outlet terminations. In the case of an outgoing trunk call, signaling information would then be transmitted to the distant switch.

There are several types of common-control systems, but two basic methods are distributed and centralized control. In many cases, the selection indicates the use of a wired-logic or stored-program approach, but wired-logic systems can be centralized as well as distributed. Both electromechanical and electronic circuits may be used in the design, but the speed normally required in centralized control, and the techniques usually employed in a stored-program approach, imply electronic components.

5.1.1 Distributed Control

Distributed-control switching systems are those that physically and functionally distribute or separate the common equipment that controls the system. They differ from centralized systems in that they use modular, or building block, hardware construction. All systems have some distribution of control elements—none uses a single entity for total and absolute control. It is the extent of relative distribution that identifies a distributed system. An additional distinguishing characteristic is the relative independence of different elements of control. In general, distributed-control systems use stage-by-stage link selection techniques.

The functional groups that comprise a typical distributed-control system are line and trunk supervisory equipment, sender and receiver common equipment, matrix-operation equipment, and call-translation equipment. Each of these groups operates somewhat independently and functions autonomously. As a result, the equipment allocated to each group requires only enough sophistication to meet the needs of that function, and call completion is accomplished by passing the control of the call through the system.

5.1.2 Centralized Control

In contrast to distributed control, *centralized control* has a master entity that has control of the call at all times. Information may be received or transmitted, matrix connections established, and other activities performed by interface equipment, but only upon direction of and under control of the central processor unit (CPU). Such a method of operation imposes a large real-time requirement which, for modest-sized systems, requires the speed of electronics. Typically, these sytems are stored-program control, employing redundant digital computers as the CPU. In larger systems, it is often necessary to provide more than one CPU to meet the requirements imposed on the system by the users. Such multiprocessor systems may be functionally or physically divided to portions of the switching matrix. In either case, some time is required to coordinate the processors.

5.2 WIRED-LOGIC CONTROL

The term *wired-logic control* refers to the intelligence and memory aspects of the common-control equipment. In a wired-logic machine the memory is essentially wired into the equipment when it is manufactured.

This requires a functional circuit designed for each operation the system must perform. These functional circuits are interconnected and are associated with a master clock such that a particular set of input signals generates an appropriate sequence of events that perform the necessary operations. Action takes place in a sequential mode which is part of the system design, and is controlled by predetermined timing signals and the exchange of signals between components. The interface equipment usually associated with wired-logic systems is discussed first, followed by a discussion of control elements.

Interface equipment serves both technical and functional purposes. From the technical point of view, interface equipment is necessary to convert small voltages of solid-state electronic circuits to the greater voltages required by electromechanical portions of the system, such as trunk circuits or the switching matrix. Another factor requiring interface is the speed of operation—electromechanical devices being extremely slow in relation to electronic devices. There are other technical aspects, but this discussion will focus on the functional aspects of interface equipment in relation to the transmission and reception of call information, and control of the matrix.

5.2.1 Information Exchange

The senders and receivers provide a means of storing information for the call-processing elements of the system. Senders transmit signaling information to distant switches. Receivers, on the other hand, detect pulses (DP) or tones (DTMF, SF, MF) sent from subscribers or distant offices. The selected digits are determined by translation of receiver-stored information, originating class-of-service data, and matrix status indications.

There are usually a number of different receiver types in a system. Subscriber lines may interconnect to a signaling register that receives only DP signals, or only DTMF signals, or both DP and DTMF signals. These receivers isolate the switching matrix from the subscriber lines, thereby removing the need to provide functional or technical compatibility between the subscriber's line and the matrix. This also permits the efficient introduction of new services. Receivers offer the ability to add or delete digits, thus enabling the common-control system to perform alternate routing.

5.2.2 Supervisory Equipment

Supervisory circuits are required for detecting changes in the state of lines or trunks on the switching matrix. These circuits fulfill the functional responsibilities for alerting, attending, and supervising. They may also provide tones and be used in maintenance or test routines.

A *line scanner* may typically consist of two scanning circuits serving one hundred lines. One scanner will have ten steps and scan leads that represent multiples of ten lines. When a request for service is detected in the group of ten, the second scanner will determine the unit digit, or one of the ten, which is requesting the service. Thus, two ten-step scanners can identify one in one hundred lines.

5.2.3 Matrix Control

The term *marker* generally refers to the device that operates upon the matrix and marks the path to be established. In a centralized system the marker will control all stages of selection, whereas in a distributed or stage-by-stage operating system there may be two or more markers to complete the connection.

Wired-logic systems do not use a full memory of matrix states, but either rely on direct path interrogation by the marker, or use an end-to-end marking technique. A direct path interrogation method requires leads in the matrix that permit testing the link states (busy or idle). The marker then operates relays on selected switch arrays that place these leads on a common data bus from the matrix. In the event there are parallel paths and more than one are idle, a selection process must take place. The link selection strategy can have a dramatic effect on matrix congestion and the permissible link loading of the matrix. Sophisticated selection strategies, however, require more information and time to collect the data and to exercise more intelligence.

Both electronic and electromechanical markers are currently in use. The electronic markers have a decided advantage in speed of operation, but require additional interface equipment to operate on the matrix. In a distributed-control system the marker will be the heart of the system, whereas in a centralized-control system the markers perform in a slave capacity to the CPU.

5.2.4 Common-Control Equipment

In wired-logic and distributed-control systems, it is difficult to identify common-control equipment or to distinguish it from interface equipment. Common-control equipment comprises those devices that exhibit the intelligence of the system and provide executive control. In a wired-logic system, those elements are the translation equipment and call-processing control.

Before any originating or incoming call can be connected through the matrix, a decision must be made concerning what is to be done with it. The translator fulfills this function in a wired-logic system. The input consists of receiver-stored information, class of service marks, line or trunk-group status indicators, and a translation memory. The output consists of instructions to markers concerning matrix connections, and to senders concerning signaling information.

There have been several types of wired-logic translators developed over the years, such as relay, ring-core, optical, and electronic. They have used space-divided and time-divided techniques and required a varying amount of effort for reprogramming. Changing the instructions is relatively complex; consequently, the flexibility of the system is inherently limited. The cost of these changes, however, may be offset by the reduced cost of memories and often proves to be an attractive alternative in small offices where changes are less frequent.

The purpose of the master-control circuit is to control the interchange of information between common control and interface circuitry and to coordinate the activities of device groups. This function may consist of a separate circuit, may be an inherent part of the translator, or may be fragmented into several other devices such as markers. The tasks involved are overseeing operations, implementing corrective procedures in the event of trouble, and implementing overload controls when required.

5.3 STORED-PROGRAM CONTROL

Stored-program control represents the latest step in the evolution of common-control switching. This might normally be associated only with centralized control, but there are systems using a stored-program CPU in a distributed-control operation. In the latter case, the CPU delegates responsibilities to wired-logic subsystems to minimize centralized memory and reduce real-time requirements. In a centralized-control environment, the CPU must be powerful and have a substantial amount of memory.

As with wired-logic control, required interface equipment has technical as well as functional responsibilities. Technical responsibilities for stored-program control are generally the same as for wired-logic, but functional responsibilities are reduced in the centralized mode of stored-program control. In a centralized-control system, digits received are stored directly in the CPU memory. The dialed digits may be read directly from line or trunk circuits via interface circuitry, or they may be received and stored in a register. Distributed-control systems use signaling registers (receivers) for complete registration of incoming digits and pretranslation functions.

5.3.1 Supervisory Equipment

Supervisory equipment in a stored-program environment consists of scanners and multiplexers to efficiently identify changes in the state of line and trunk circuits and communicate this information to the CPU. These signals are read at specified intervals and identify the requirements for program routines.

5.3.2 Matrix Control

In a distributed-control application, the stored-program system may leave path selection to the markers. In this case, the marker in a stored-program system may not differ substantially from the marker in a wired-logic system. There are some different techniques that can be used with a stored-program system, however. The ultimate in centralized control uses a mapping technique that creates a memory map of the status of every link, inlet, outlet, and service circuit. In this manner, the complete connection path can be selected by the CPU at very high speed, replacing the matrix interrogator and path selection functions of the wired-logic marker. This method leads to a very efficient use of the matrix, but requires a large amount of memory and complex programming of the work routines. With today's processor speeds and memory capacities, this is not a significant system constraint.

Another factor of matrix control is the latching mechanism used by the crosspoint device. Some crosspoints are self-releasing; that is, when the called subscriber hangs up, the switching matrix is released. These devices are electrically operated and held, although the crosspoints may be electromechanically latched. There are other crosspoints that are physically latched, and require no power while in use for a connection. These mechanisms have one major drawback, and that is the release—they must be disconnected during a release cycle, just as they were connected during the call-setup cycle. Thus, there are two matrix operations per call instead of one. Stored-program control and matrix mapping make this feasible because a word in memory can be used to store the path associated with a call.

A third factor of matrix control is the possibility of rearranging live calls on the matrix to enable a new call to be completed. The principle is to essentially eliminate idle-link mismatch by rearranging the matrix paths. Again, complete knowledge of the matrix status is required and considerable real time may be necessary to make such changes wisely.

5.3.3 Input-Output Devices

Input-output (I-O) devices are required to communicate with the computer's electrically alterable memory. Typically, these are visual display units (VDU) in a switching system with disk storage. This permits off-line preparation with high-speed entry into the CPU. One advantage of these I-O devices is that they can be remotely located, providing improved flexibility in making updates or controlling the system on a real-time basis.

5.3.4 Common-Control Equipment

A stored-program control system is built of hardware and software. The hardware is the collection of logic circuits that constitute the processor and perform certain basic functions as dictated by the software. The elements of software are permanent instructions and alterable data. Instructions are stored in memory and direct the computer to operate on data or to control the system hardware. These instructions are used to command the processing units to perform simple processes such as additions, subtractions, and comparisons on the appropriate data. Because the stored instructions can be changed with relative ease, the operation of the system can be changed when required without changing any of the circuitry. From a technical point of view, all that is required is that the new series of instructions be prepared and entered into the memory.

The computer must be able to handle many calls simultaneously; therefore, the call-processing programs are designed to be time shared by many subscribers. This is accomplished by breaking the processing of each call into a series of stages each of which, when executed, advances the call one step closer to completion. Thus, the processor is time shared in the sense that it attempts to sequentially process one step of each call in the system rather than attempting to process all stages of one specific call. All programs in a stored-program system may be categorized by the functions they perform. These are divided into executive programs, directive programs, and work routines.

Executive programs are provided on various priority levels. They ensure proper entry to the priority level, supervise work contained in the directive programs, and ensure proper exit from the priority level. Executive programs work with a special file, called a job table, on those levels controlled by a periodic interrupt signal. The job table contains the address of each directive program and its recall interval. Scanning the job table at a given time indicates which program should be recalled at that time.

Directive programs indicate the work to be done by the processing subsystem, whereas *work routines* actually perform the required tasks. Note that call processing differs substantially from data processing. The difference lies in that call processing involves real time, whereas data processing does not. An error in the program of a stored-program system could result in a catastrophic system failure, depriving several thousands of subscribers of service. Because work in a switching system must be carried out in real time, the work programs are organized on priority levels, with transfer to a higher level occurring when required.

The highest levels are reserved for supervision and fault elimination, and the remaining levels are used for less urgent functions. An increase in priority level, that is, changeover to a program on a higher priority level, is initiated by one of the following types of interrupt signal:

a. clock interrupt (periodically repeated) signals;
b. externally initiated interrupt signals, such as an interrupt signal to make a transfer of data from an external device;
c. Internally initiated interrupt signals, such as an interrupt signal when a system fault is detected.

Upon receiving an interrupt signal, a change in the priority level will occur at that time only if the interrupting level is of a higher priority than the interrupted level. When this is the case, the register contents and return address to the current program are stored. Thus, when the higher priority task is completed, the program may resume at precisely the point at which it was interrupted.

5.4 COMMON-CONTROL CONGESTION ANALYSIS

In a S × S system, equipment is operated as dialing occurs and each dialed digit is lost once it has performed its path selection function. However, in a common-control system, the switching matrix is unable to respond to dialed information and the matrix connections are completely under control of the common equipment. The result is that the effectiveness of the switching system is very dependent upon the efficiency of the common-control equipment. Common control results in greater flexibility and efficiency in switching, but introduces complexities that require a careful analysis of the entire switching system.

5.4.1 Areas of Congestion

In addition to the switching-matrix congestion objectives, a common-control system must define the objectives for access to the system and objectives for call disposition or speed of switching. The first type of objective deals with the initial area of potential congestion a call encounters—access to the receivers in order to transmit information to the system. The second type of objective concerns the speed with which the system switches the call once all the necessary information has been received.

Both of these areas of potential congestion deal with an element of delay rather than simply the blocking probability. Because the holding times on common-control devices are relatively short, the call can often wait until a server becomes idle rather than being cleared from the system immediately, as in the case of the switching matrix. The objectives, then, are expressed as a probability of being delayed for longer than some specified time, and are selected on the basis of a tolerable waiting time by the subscriber, or the effect on other control elements in the system.

5.4.2 Receiver Attachment

There are a number of specific objectives that must be met in order to provide adequate access to the system. Incoming trunks require an objective for either controlled-access delay or for interdigital-access delay. Controlled-access trunks are incoming from other common-control offices and will wait until a receiver has been attached and a start-sending signal has been returned before they begin to send address information. The controlled-access delay objectives offer no real problem to the system because these objectives are on the order of five in a thousand (0.005) delayed over three seconds. It should be noted, however, that this delay is a composite delay distribution of interface equipment, connecting links, control, and other items involved in the connection.

5.4.3 Interdigital-Access Delay

Interdigital-access trunks are incoming from S × S offices and are connected directly through to the subscriber's dial. Obviously, the subscriber does not wait for a receiver to be attached in the middle of dialing, but continues on, unaware of any switching that is taking place. The interdigital-access trunks are most critical since very little delay can be tolerated. The subscriber must be connected to a DP receiver between digits, and a relatively large amount of interdigital time is consumed by the originating S × S office. The result of a late connection can be a misrouted call (when one or more pulses are missed in a digit), or an incomplete registration of digits (when an entire digit is missed); therefore, objectives for these connections are typically on the order of one in a thousand (0.001) being delayed longer than 50 milliseconds (ms).

By the time the distant outgoing selector has recognized the inter-digital pause, started rotary, located an idle bank contact, and seized the outgoing trunk, most of the interdigital pause is gone, leaving very little switching time available for the common-control system to recognize the request for service and attach a receiver. This is a real problem with fast dialers, which studies have shown offer pauses as short as 300 ms.

One solution to this problem would be a second dial tone, but the subscriber might not know which digit to wait for; therefore, such an arrangement does not offer acceptable service. A second approach is an extremely fast connect time with very little probability of delay. In some systems a bypath matrix is provided for such trunks to establish a very high speed connection (on the order of 50 ms). A third approach is to store a pulse of the next digit in the trunk circuit. At a standard ten PPS dial speed, this extends the interdigital time by 100 ms. The principle is to store the first pulse of the first digit received at the system in the trunk circuit, and then add a pulse to the digit stored in the receiver. A fourth approach used by some stored-program systems is to read the pulses directly from the trunk circuit by scanning the status indicators.

5.4.4 Dial-Tone Delay

Another objective is only required at COs, and is concerned with dial-tone delay for originating calls. In this case, subscribers normally wait until a dial tone is heard before dialing. As with controlled-access delay, dial-tone delay objectives are relatively relaxed. The industry standard is fifteen in a thousand (0.015) delayed over three seconds. Again, total system operations required to establish the connection should be included in the analysis of this objective. This may involve several operations in a wired-logic system, including the matrix connection to a receiver. In a stored-program system, such a connection may be required for DTMF subscribers, but not for DP subscribers if the dial pulses are read from the line circuit and accumulated in memory.

5.4.5 Speed of Switching

In order to keep the common-control equipment operating efficiently, objectives are established that ensure that a good flow of calls can be maintained. Common-control systems require an even exchange of infor-mation between the functional subsystems of a wired-logic system, or require a controllable real-time load on a stored-program system. Typi-cally, objectives are established that no more than one call in a thousand

(0.001) of all calls are delayed longer than one second (including the operate time of control and matrix components) in switching. In some systems, translators may be time divided, in which case there is no delay. In wired-logic systems translators as well as control circuits, markers, and other related devices of the system may cause delays. Stored-program systems require that the CPU be able to operate at relatively low priority levels most of the time, minimizing priority interrupts which take real-time capacity from switching operations.

The key to efficient common-control system operation is the elimination of overloads and minimization of delays in all entities of control. An overload in one portion of the system becomes a bottleneck that can seriously impair the efficiency of other parts of the system. For this reason, common-control systems must have efficient overload controls, referred to as network-management and switch-management tools. It is the function of these controls, operated manually or automatically, to effectively control localized overloads and prevent them from producing generalized system overloads. This has become a science in its own right, and more sophisticated techniques are being introduced all the time. Dynamic controls exist that not only change the operating parameters of the system, but also signal distant offices to implement controls and change routing so that the overloaded system may recover and regain efficiency.

5.4.6 Method of Analysis

When dealing with delay aspects of congestion, two factors must be identified that influence the evaluation. First, the nature and duration of the holding times must be known. Often, the holding time is determined by design, such as markers, control circuits, and CPU access time. In some cases the holding time is relatively constant or can be defined by a limited number of events and their probability of occurrence. In other cases the holding time has a negative-exponential distribution which can easily be treated in traffic theory.

The second factor to be considered is the service discipline, or the manner in which delayed calls are served. There are several standard techniques such as order-of-arrival (FIFO); random; or a gating method that permits groups of calls to come in and then selects randomly within the group. The gating method is a compromise between FIFO and random service disciplines, and is more convenient to design than a FIFO queue mechanism.

Once these two factors have been defined, along with source assumptions, offered load, and number of servers, standard formulas and tables may be used to determine the average delay and the delay distribution, or the probability of being delayed longer than some specified time (see Chapter 3). These formulas and tables are based on a critical examination that assures that the engineering criteria, together with the design factors, will meet the overall objectives. Such an analysis often requires an evaluation of tandem queues, in which case more than one device in sequence can cause delay, and thus contribute to the overall delay objective. Such evaluations require convolutions of discrete probability distributions, and are beyond the scope of this book. An alternative and widely used technique is computer simulation.

Stored-program systems require a highly detailed analysis of the functions to be performed by the system and the relative real-time requirement to be imposed on the CPU. This requires the definition of each function that has any possibility of occurrence, and the frequency with which it will occur. Such evaluations are difficult due to the interrelationship of occurrences, and accurate estimates of real-time capacities of stored-program systems are often dependent upon simulations and field experience.

5.5 PROCESSOR-LOADING ANALYSIS

The CPU has finite limits on its capacity to process real-time traffic, plus peak-traffic loading. One of the key parameters of concern is the CPU operating speed, specified in instructions per second (inst/s). The more instructions a CPU can process in real time, the less likely it will become overloaded under heavy traffic conditions. It is important to note that any switching system, even one with a nonblocking matrix, will encounter service degradation if the CPU cannot handle the offered traffic load.

Consider an international gateway exchange using CCITT #5 signaling on the international side and CCITT R2 signaling on the national side. The following system parameters are applicable for this example:

CPU Speed: 300,000 inst/s
Offered Traffic: 1200 Erlangs during busy hour
Call-Holding Time: 180 seconds (average)
Number of Digits: 14 for international calls (both ways)
           14 for calls from the national exchange
           8 for calls to the national exchange
Number of Registers: 10 CCITT #5 (both ways)
           12 CCITT R2 (incoming)
           7 CCITT R2 (outgoing)
Serial-Line Interfaces: 10 (between CPU and line processors)

5.5.1 Fixed-Processing Tasks

Fixed-processing tasks are those that are performed regardless of whether there is traffic. Examples of this type of task are signaling register scanning, real-time clock functions, database updates, and on-line diagnostic routines. Assume that the fixed-processing tasks for this example have the following parameters:

Clock Interrupts:	10 serial-line interfaces (SLI)
	2 messages per SLI
	8 instructions per message
	20 interrupts per second
Register Scanning:	29 registers
	16 instructions per register
	40 interrupts per second
Database Update:	60 entries (average)
	9 instructions per entry
	20 interrupts per second
Miscellaneous:	40 instructions per interrupt
	20 interrupts per second

The total fixed-processing load is the sum of the loading contributed by each fixed-processing task. Therefore, the fixed-processing task loading may be calculated as follows:

Clock Interrupts:	(10)(2)(8)(20) =	3,200 inst/s
Register Scanning:	(29)(16)(40) =	18,560
Database Update:	(60)(9)(20) =	10,800
Miscellaneous:	(40)(20) =	800
Total Fixed Load:		33,360 inst/s

5.5.2 Variable-Processing Tasks

Variable-processing tasks are those that are performed during call processing, and are a function of the switch traffic. Examples of this type of task are executive functions, signal processing, number translation, and message processing. Assume that the variable-processing tasks for this example have the following parameters:

Executive Functions:	3 scans per call
	1200 instructions per scan
Incoming Signals:	86 interrupts per call
	40 instructions per interrupt

Outgoing Signals:	118 interrupts per call
	45 instructions per interrupt
Number Translation:	60% of traffic is CCITT #5 (both ways)
	14 digits per call
	25% of traffic is CCITT R2 (incoming)
	14 digits per incoming call
	15% of traffic is CCITT R2 (outgoing)
	8 digits per outgoing call
	800 instructions per translation
Message Processing:	16 messages per call
	300 instructions per message
	28 acknowledgments per call
	80 instructions per acknowledgment

The total variable-processing load is the sum of the loading contributed by each variable-processing task. Therefore, the variable-processing task loading per call may be calculated as follows:

Executive Functions:	$(3)(1200) =$	3,600 inst
Signal Processing:	$(86)(40) =$	3,440
	$(118)(45) =$	5,310
Number Translation:	$(0.6)(14)(800) =$	6,720
	$(0.25)(14)(800) =$	2,800
	$(0.15)(8)(800) =$	960
Message Processing:	$(16)(300) =$	4,800
	$(28)(80) =$	2,240
Total Variable Load per Call:		29,860 inst

5.5.3 Processor-Load Factor

The processor-load factor (PLF) is the ratio of the busy-hour loading to the CPU capacity and may be determined using (5.1). In order to handle the busy-hour plus peak loading, the PLF must be less than one with a reasonable (15 percent or more) margin.

$$\text{PLF} = [(\text{Fixed Load}) + (\text{Variable Load})]/(\text{CPU Capacity}) \quad (5.1)$$

$$\text{PLF} = [33,360 + (29,860)(N)]/(300,000)$$

where

$$N = (\text{Offered Traffic/Call-Holding Time}) = (1200 \text{ Erl.}/180 \text{ s}) \approx 6.7 \text{ calls/s}$$

$$\text{PLF} \approx [33,360 + (29,860)(6.7)]/(300,000) \approx 0.78 \ (22\% \text{ margin})$$

Chapter 6
Precedence and Preemption

The purpose of this chapter is to present a quantitative means for determining the probability that subscribers, interswitch trunks, or servers may be preempted from an existing call to serve a higher-precedence call attempt. Emphasis is placed on the practical application of the preemption-probability equations rather than on their derivation.

6.1 MULTILEVEL-PRECEDENCE SYSTEMS

Each subscriber in a multilevel-precedence system is granted a maximum precedence level that, except for the lowest level, may be invoked to improve the probability that a call attempt will be completed. A new call attempt may not preempt (replace) any existing call protected at an equal or higher-precedence level, and calls at the lowest-precedence level will always be preempted first. Therefore, a call initiated at a high level will be reasonably protected from preemption. However, a call initiated at the lowest level cannot preempt any other call, and is subject to preemption by any precedence call.

Precedence and preemption, then, improve call completion probability for high-precedence subscribers at the expense of degraded service for lower-precedence subscribers. In addition, the effects of precedence and preemption include changes in the traffic pattern. Preempted calls are cut short, leading to reduced mean call-holding time. Preempted subscribers attempt to reinitiate calls, thus increasing the call-attempt rate and common-equipment loading. These factors, plus the precedence and preemption processing requirements, increase loading on the CPU.

6.1.1 Applicable Assumptions

The preemption-probability equations are based on the assumptions of random call-arrival rate; negative-exponential call-holding time; call arrivals independent of subscribers' precedence levels; full access to all trunks in the trunk group; preemption probabilities independent of each other; random preemptions; friendly-preemption discipline; and no special communities of interest.

6.1.2 Call-Precedence Distribution

The probability of preemption for a call protected at any given precedence level is dependent on the distribution of dialed precedence at call initiation. Let the call-precedence distribution be

$$Y_1, Y_2, \ldots, Y_N$$

where

Y_1 = Distribution for the highest-precedence level

Y_2 = Distribution for the next highest-precedence

level

Y_N = Distribution for the lowest-precedence level

$$Y_1 + Y_2 + \ldots + Y_N = 1$$

Then, let ΣY_j represent the summation of precedence distributions higher than Y_i. If call-precedence distribution is not specified, assume (for a worst-case analysis) that subscribers initiate all calls at their maximum authorized precedence level. For example, a two-level precedence plan (priority and routine) yields

$$\Sigma Y_1 = 0$$

$$\Sigma Y_2 = Y_1$$

That is, if 10% of the calls are initiated at the priority (P) level, then 90% of the calls are initiated at the routine (R) level ($\Sigma Y_2 = 0.1$). Similarly, if 30% of the calls are initiated at the P level, then 70% of the calls are initiated at the R level ($\Sigma Y_2 = 0.3$). Any call initiated at the P level may preempt an existing call initiated at the R level. Obviously, the lower the percentage of high-precedence calls, the lower the probability of preemption of low-precedence calls. This reasoning may be extended to any precedence level. Note that calls protected at the highest-precedence level will never be preempted ($\Sigma Y_1 = 0$).

6.2 PREEMPTION PROBABILITY

Preemption probability is dependent on the congestion probability for the various classes of service. That is, their congestion predictions must be multiplied by the appropriate preemption-probability factor to determine the probability that those resources will be preempted.

6.2.1 Subscriber-Line Preemption

A subscriber line involved in an established call will be preempted if a higher-precedence call is directed to either of the two subscribers in the connection. The probability of a call being so directed is given by the factor $2NA/(N - 1)$; where N is the number of subscribers on the switch, and A is the average subscriber traffic (Erlangs per subscriber). Therefore, the probability that a subscriber line involved in a call protected at a given precedence level will be preempted may be calculated as follows:

$$P_L(Y_i) = [2NA/(N - 1)]\Sigma Y_j \tag{6.1}$$

Note that as N increases, preemption probability rapidly becomes independent of its value because $N/(N - 1)$ approaches unity. Therefore, when N is large subscriber-line preemption probability is approximately

$$P_L(Y_i) \approx 2A\Sigma Y_j \tag{6.2}$$

6.2.2 Trunk Preemption

A friendly trunk-preemption discipline is one wherein a high-precedence call arriving at a trunk group with all trunks busy searches for an idle trunk in all alternate routes before initiating a preemptive search of the primary trunk group. Conversely, a ruthless preemption discipline is one wherein the blocked high-precedence call will immediately initiate a preemptive search of the primary trunk group (Calabrese et al. 1980). The friendly discipline takes longer and requires more processing, but results in fewer preemptions (higher probability of finding an idle trunk).

The probability that an idle trunk is not available (trunk-group grade of service) is given by the Poisson or Erlang-B distribution, as applicable. The probability that at least one trunk is busy at a lower-precedence level is given by the factor $(1 - \Sigma Y_j^N)$; where N is the average trunk-group size. Therefore, the probability that a trunk protected at a given precedence level will be preempted may be calculated as follows:

$$P_T(Y_i) = P(1 - \Sigma Y_j^N) \Sigma Y_j \qquad (6.3)$$

where

P = Trunk-group grade of service (Poisson or Erlang-B)

N = Average number of trunks in trunk group

Note that as N increases, preemption probability becomes independent of its value because $(1 - \Sigma Y_j^N)$ approaches unity. Therefore, when N is large trunk-preemption probability is approximately

$$P_T(Y_i) \approx P \Sigma Y_j \qquad (6.4)$$

Because it is possible that more than one interswitch trunk group may be involved in a network call involving a number of tandem switches, the composite probability of trunk preemption may be calculated as follows:

$$P_T(Y_i) = 1 - [1 - P_{Ta}(Y_i)][1 - P_{Tb}(Y_i)] \ldots [1 - P_{Tn}(Y_i)] \qquad (6.5)$$

where

$P_{Ta}(Y_i)$ = Probability of preemption in group a

$P_{Tb}(Y_i)$ = Probability of preemption in group b

$P_{Tn}(Y_i)$ = Probability of preemption in group n

6.2.3 Server Preemption

Pooled servers, such as signaling registers, are provided in a switching system to serve all call requests. They may be preempted if required to process a higher-precedence call, and no other server of the same class is idle or protected at a lower-precedence level. Separate queues are provided for each precedence level, and calls waiting for service are taken in order from higher-precedence queues first. The probability that an appropriate server is not available (server-pool grade of service) is given by the Erlang-C distribution. The probability that at least one server is busy at a lower-precedence level is given by the factor $(1 - \Sigma Y_j^N)$; where N is the number of servers in the pool. Therefore, the probability that a server involved in a call protected at a given precedence level will be preempted may be calculated as follows:

$$P_S(Y_i) = P(> 0)(1 - \Sigma Y_j^N)\Sigma Y_j \qquad (6.6)$$

where

$$P(> 0) = \text{Server grade of service (Erlang-C)}$$

Note that as N increases, preemption probability becomes independent of its value because $(1 - \Sigma Y_j^N)$ approaches unity. Therefore, when N is large server-preemption probability is approximately

$$P_S(Y_i) \approx P(> 0)\Sigma Y_j \qquad (6.7)$$

Because more than one class of server may be involved in a given call, particularly during the signaling phase of a network call involving a number of tandem switches, the composite probability of server preemption may be calculated as follows:

$$P_S(Y_i) = 1 - [1 - P_{Sa}(Y_i)][1 - P_{Sb}(Y_i)] \ldots [1 - P_{Sn}(Y_i)] \qquad (6.8)$$

where

$$P_{Sa}(Y_i) = \text{Preemption probability for equipment class } a$$

$$P_{Sb}(Y_i) = \text{Preemption probability for equipment class } b$$

$$P_{Sn}(Y_i) = \text{Preemption probability for equipment class } n$$

6.2.4 Call Preemption

A network call may be interrupted due to the preemption of either the calling or called subscriber lines, one or more interswitch trunks, or one or more servers anywhere in the network. Therefore, the probability that a call protected at a given precedence level will be preempted may be calculated as follows:

$$P_N(Y_i) = 1 - [1 - P_L(Y_i)][1 - P_T(Y_i)][1 - P_S(Y_i)] \qquad (6.9)$$

where

$P_L(Y_i)$ = Subscriber-line preemption probability

$P_T(Y_i)$ = Composite trunk-preemption probability

$P_S(Y_i)$ = Composite server-preemption probability

6.3 FIVE-LEVEL PRECEDENCE EXAMPLE

Consider a military AUTOVON (Automatic Voice Network) switch using five levels of precedence (flash override, flash, immediate, priority, and routine) as a model to demonstrate the use of the equations. Assume the following system parameters for this model:

a. busy-hour traffic of 0.2 Erlangs per subscriber;
b. 500 subscriber lines per switch;
c. average trunk group of 20 trunks;
d. trunk-group grade of service of $P = 0.01$;
e. average common-equipment pool of 10 servers;
f. common-equipment grade of service of $P(> 0) = 0.01$;
g. call-precedence distribution per Table 6.1.

Table 6.1 Call-Precedence Distribution

Precedence Level	Y_i	Percent	ΣY_j
Flash Override	Y_1	0.2	—
Flash	Y_2	1.8	0.002
Immediate	Y_3	4.0	0.020
Priority	Y_4	28.0	0.060
Routine	Y_5	66.0	0.340

6.3.1 Subscriber-Line Preemption

Equation (6.1) may be used to determine subscriber-line preemption probabilities for each precedence level as follows:

$$P_L(Y_i) = [2NA/(N - 1)] \Sigma Y_j$$
$$= [(2)(500)(0.2)/(499)]\Sigma Y_j \approx (0.4)\Sigma Y_j$$

$$P_L(Y_2) \approx (0.4)(0.002) = 0.0008$$
$$P_L(Y_3) \approx (0.4)(0.02) = 0.008$$
$$P_L(Y_4) \approx (0.4)(0.06) = 0.024$$
$$P_L(Y_5) \approx (0.4)(0.34) = 0.136$$

6.3.2 Trunk Preemption

Equation (6.3) may be used to determine trunk preemption probabilities for each precedence level as follows:

$$P_T(Y_i) = P(1 - \Sigma Y_j^N)\Sigma Y_j = (0.01)(1 - \Sigma Y_j^{20})\Sigma Y_j \approx (0.01)\Sigma Y_j$$

$$P_T(Y_2) = (0.01)[1 - (0.002)^{20}](0.002) \approx 0.00002$$
$$P_T(Y_3) = (0.01)[1 - (0.02)^{20}](0.02) \approx 0.0002$$
$$P_T(Y_4) = (0.01)[1 - (0.06)^{20}] (0.06) \approx 0.0006$$
$$P_T(Y_5) = (0.01)[1 - (0.34)^{20}] (0.34) \approx 0.0034$$

6.3.3 Server Preemption

Similarly, Equation (6.6) may be used to determine server-preemption probability for each precedence level as follows:

$$P_S(Y_i) = P(> 0)(1 - \Sigma Y_j^N)\Sigma Y_j = (0.01)(1 - \Sigma Y_j^{10})\Sigma Y_j$$

$$P_S(Y_2) = (0.01)[1 - (0.002)^{10}](0.002) \approx 0.00002$$
$$P_S(Y_3) = (0.01)[1 - (0.02)^{10}](0.02) \approx 0.0002$$
$$P_S(Y_4) = (0.01)[1 - (0.06)^{10}](0.06) \approx 0.0006$$
$$P_S(Y_5) = (0.01)[1 - (0.34)^{10}](0.34) \approx 0.0034$$

6.3.4 Call Preemption

Equation (6.9) may be used to determine call-preemption probabilities for each precedence level as follows:

$$P_N(Y_i) = 1 - [1 - P_L(Y_i)][1 - P_T(Y_j)][1 - P_S(Y_j)]$$

$$P_N(Y_2) = 1 - (1 - 0.0008)(1 - 0.00002)(1 - 0.00002) = 0.00084$$

$$P_N(Y_3) = 1 - (1 - 0.008)(1 - 0.0002)(1 - 0.0002) = 0.00839$$

$$P_N(Y_4) = 1 - (1 - 0.024)(1 - 0.0006)(1 - 0.0006) = 0.02465$$

$$P_N(Y_5) = 1 - (1 - 0.136)(1 - 0.0034)(1 - 0.0034) = 0.14187$$

REFERENCE

Calabrese, D.A., M.J. Fischer, B.E. Hoiem, and E.P. Kaiser, "Modeling a Voice Network with Preemption," *IEEE Transactions on Communications*, Vol. COM-28, No. 1, January 1980, p. 22.

Chapter 7
Planning and Forecasting

There was a time when the responsibilities of traffic engineers were heavily weighted in favor of manual operations required to complete calls. This is still an important function of teletraffic engineering; however, the concept of direct responsibility toward the planning, engineering, and administration of all switched services has materially changed the content of our duties. We now have responsibilities in many areas which heretofore scarcely received much of our consideration or time. We also contribute to plans for new techniques employing electronic switching and ticketing systems for local and toll services, and new techniques for doing the daily work of traffic engineering with improved efficiency through the use of computers.

7.1 FACILITY PLANNING

Facilities are provided for one basic purpose—to permit people to talk to people and machines to talk to machines. In short, facilities are provided to carry traffic. Planning for facilities involves selecting the optimum type of facility to meet a specific need, and the determination of the best arrangement of these facilities. All types of facilities are considered, including local switching, toll switching, and ticketing equipment, plus the interoffice trunks required to tie these units together. Even land, buildings, and power subsystems must be sized, at least in part, on the basis of traffic information.

7.1.1 Types of Plans

The telephone plant construction process consists of fundamental planning, current planning, and then implementation of the approved plans. The quality and efficiency of service is dependent on the selection of the optimum switching system as determined by the fundamental plan, on the optimum arrangements selected as part of the current plan, and on the proper sizing and timing of the project as defined by the implementation program. Traffic engineers contribute to all three.

Fundamental planning is the process of determining the type of facility and basic arrangements of the facility. In this phase only those details necessary to make the basic decisions are considered. The service life of most facilities should exceed twenty years; therefore, these decisions should be based on the best view of that long-range period.

Current planning expands on the framework of the fundamental plan. There are usually many problems and alternative arrangements that must be studied after the decision has been reached to install a particular system. These options are not of sufficient importance to alter the fundamental plan, but may require economic selection studies to determine optimum arrangements and timing. Current planning should stress a shorter time interval than fundamental planning, but should also consider several engineering periods to ensure smooth growth with minimum rearrangement and changes.

7.1.2 Engineering Specifications

Planning is a cooperative effort—every department, including management and administration, has an interest in the results and contributes to the process. Therefore, a reliable means of disseminating information to the other departments is required. The traffic order (TO) is used to transmit the detailed information required for the equipment engineering and implementation phases.

In order to facilitate the use of the TO by several groups within a company, a uniform format is recommended. The TO should contain general notes, method of operation, fundamental data, basic traffic data, equipment details, summary of equipment, and equipment diagrams that fully describe the necessary details. In addition, documentation should be included to support decisions reached during the traffic system design effort. If not included in the TO, there should be a definite method whereby such notes are retained for reference.

The implementation phase is the final step in any plan. The timing and extent of implementing the current plan is constrained by the construction budget. Once a project is defined by the TO, close cooperation is required between traffic engineering, equipment engineering, and plant engineering to achieve the ultimate goal of equipment in service. Traffic engineering prepares the TO, equipment engineering prepares the detailed equipment specification, and plant engineering is responsible for construction of the facility.

7.1.3 Planning Process

In order to illustrate some of traffic engineering's responsibilities, assume that a decision has been reached to establish a new local office to serve a defined area. The problem then is to select the type of switching equipment that will best serve that area, as detailed by the following steps.

Analyze the lines and main stations presently served in the area based on class of service. List the services and features that are presently available. Delineate the present serving arrangements and determine the characteristics of the traffic generated by those subscribers. These characteristics include calling rates, busy hours, communities of interest, *et cetera*. Forecast and analyze the requirements of the new office.

The forecast of services should contain all services that will be, probably will be, or may be provided. The forecast of lines and main stations will determine the number of terminations that must be provided. The forecast of traffic characteristics must reflect existing characteristics that have been increased according to trends, simulated for new services, and adjusted in light of any changes in the character or economics of the area.

7.2 TRAFFIC FORECASTS

Traffic engineering provides a large portion of the framework for any study of telephone plant construction. During fundamental planning, the requirements of each specific application must be forecast for comparison with the capabilities of different systems. Current planning requires similar forecasts in order to evaluate alternative arrangements.

Traffic engineers must prepare traffic forecasts for periods ranging up to twenty years. These should include statistical data plus originating and terminating traffic based on class of service or logical equipment types. Calling rates have been accelerating for many years, and they will most likely continue to increase. They may even be stimulated by new service offerings or may change with new types of traffic, such as optional dialing plans, data services, *et cetera*. The traffic engineering department is the organization best equipped for the study of such trends.

7.2.1   Traffic Volume and Distribution

Traffic engineering's stock in trade is the flow of calls through the switched network. For example, we are aware of the interactions between switching units within the network. This is the type of insight that most nearly matches the processes involved in planning. This same insight is also invaluable for determining the detailed arrangement of equipment most likely to prove acceptable.

Traffic volumes are required because the facility must be capable of carrying the traffic generated. The capacity and range must permit economical operation from the initial installation until replacement. The sizing of all switching, ticketing, and trunking facilities is determined by traffic volumes.

The distribution of traffic must also be determined. It is not sufficient just to specify the number of trunks in a group—its length and connections at both ends must also be specified. The distribution of telephone calls is needed to properly plan the routing patterns, determine the required trunk groups, and size each group. Therefore, the traffic distribution and other characteristics of calls are constantly under study.

Alternative systems that can meet the basic requirements must be determined. The next step is to compare the system requirements with the capabilities of each alternative. The objective is to eliminate from consideration any system or subsystem that cannot meet the requirements. One comparison to be made is the services required *versus* the features available. This comparison may not only eliminate some of the alternatives, but may also reveal optional features that must be considered in the economic selection studies.

Traffic load *versus* traffic capacity, and the ability to provide for unforeseen loads or services, must be compared. Will each of the alternative systems carry the anticipated load with equivalent efficiency? When two or more options remain, comparative forecasts that are adapted to each type of equipment must be prepared. Contingency forecasts using alternative assumptions may reveal inefficiencies in some systems.

7.2.2   Comparative and Contingency Forecasting

Comparative forecasting considers each system individually. Because the objective is to select between alternative systems or arrangements, forecasts must be prepared that are tailored to each alternative. Each forecast must consider comparable grades of service and traffic volumes.

Contingency forecasting takes the questionable areas into account. There are many occasions when the future is in doubt, and there is a tendency to assume one event or another will occur, or to assume a specific time for an event to occur. Forecasts made for both cases and their extremes will identify the maximum and minimum situations, and reveal the amount of flexibility that is in the plan.

Finally, equipment engineering prepares economic selection studies (trade-off analyses). Only those alternatives that can meet all of the requirements at standard grades of service are priced out and considered for the implementation phase.

7.2.3 Noneconomic Considerations

Still another traffic engineering contribution is in the area of non-economic considerations. The obvious example that comes to mind is the question of the labor market. Are there adequate qualified people to meet anticipated needs? There may be a tendency to downgrade the importance of such considerations, but they may be the deciding factors regardless of the apparent economics. Management constantly reviews the planning process and, other things being equivalent, may authorize the selection of one system over another on the basis of noneconomic considerations.

Network considerations should reflect the effects of a new office on other offices. Is the present pattern of routing suitable or is some rearrangement required? Perhaps a complete retrunking with one or more tandem switching centers will be required. Traffic engineers are able to recognize and reveal problems that may arise in this area and can suggest solutions.

Another consideration in selecting a switching system is its flexibility and adaptability to new services. Consider a stored-program system which may only require software changes to add new services. If other factors are not predominant, this flexibility may be decisive.

7.3 FORECASTING TECHNIQUES

Forecasting for planning is not basically different from forecasting for other purposes. There are some conditions that change slowly and may be safely ignored in short-range forecasting, whereas they become more meaningful in the long-range view required for planning. An example of this is the change in community-of-interest factors due to the changing character of communities within a study area.

Forecasts prepared for planning should be compared with any other traffic forecast. Differences must be justified (underlying assumptions may cause such differences). This comparison should be made in the early stages with basic forecasts, rather than after considerable labor has been expended. Ideally, forecasts for all purposes should be made at one time and in one operation. The responsibility may be either assigned to one individual or shared. If shared, close coordination is required. Another method is to accept, after evaluation, the projection factors used in preparing the short-range forecast and add correlating factors for other years.

Comparative forecasts are required whenever more than one solution to a problem requires detailed evaluation. When parameters are questionable, contingency forecasts are required. Some forecasts are for local service and some are for toll service—different types of data are required for each. Some forecasts are for individual offices and some are for groups of interconnecting offices (networks). Here again the type of data and techniques vary. Some forecasts are for calling rates and some are for traffic volumes. In every case, however, forecasts of trunks or equipment quantities must always be preceded by a forecast of traffic volume and distribution.

An orderly approach to forecasting will minimize the effort and ensure that the end result represents the best judgment possible. The following steps are inherent in the forecasting process:

a. analysis of historical data;
b. analysis of future trends based on the historical trends;
c. selection of the planning base;
d. determination of future traffic rates, volumes, and distribution;
e. calculation of equipment and trunk quantities;
f. other considerations, such as network requirements.

7.3.1 Analysis of Historical Data

One of the prime clues to the future is available in existing traffic records. These records reveal past trends for an existing office, trunk group, or network. Although the future will be affected by other factors and past trends may change, these records are a major source of information that will guide the forecasting process. The lack of adequate and reliable historical data has been a hindrance to proper forecasting. With the advent of new services and constantly changing parameters, better record keeping and more precise analysis methods are necessary.

The method of analyzing historical records may be as simple as visually scanning graphed data. While scanning graphs, an effort should be made to establish the approximate type of trend that exists. For example, is the growth straight line as characterized by the same volume increase each year, exponential as characterized by a constant percentage increase each year, or fluctuating between the above trends due to discontinuities, economic conditions, or other factors?

Mathematical methods are available that are simple enough for manual analysis or complex enough to require the use of a computer. Even when more sophisticated techniques are used, a preliminary examination of the data is advisable. This will frequently reveal situations that may completely upset a statistical trend analysis.

The cyclical nature of traffic prevents the use of data that is not representative of busy-season conditions. Nonbusy-season data that have been adjusted to approximate busy-season volumes does not warrant sophisticated techniques of analysis because the accuracy and reliability of the data are inadequate. Even the most advanced computer cannot obtain outputs more accurate than the accuracy of its input data.

Discontinuities may occur from many causes, such as area transfers that affect either the traffic calling rates, volume, or distribution; new services; changes in methods of handling traffic; and changes in the method of obtaining and recording data. Abnormal conditions (severe storms, strikes, disasters, *et cetera*) that should not be used for engineering purposes should not be included in the base data used for forecasting.

Records of several years or more are desirable. However, these are rarely available without discontinuities or other defects. If inadequate data from a lesser period are available they may be used, but only to the extent justified. If historical data are unavailable or unusable, the planner must use good judgment based upon more generalized knowledge of similar exchanges or situations. It may even be justifiable to omit the evaluation of any historical data and place added emphasis on future trending to establish the best possible forecasting or engineering base.

Table 7.1 presents busy-season, busy-hour traffic data for a hypothetical office which may be used as an example to demonstrate data analysis techniques. The following may be concluded from a cursory examination of the data:

a. The annual increase in total originated CCS/MS fluctuates—the highest increase (0.11) is followed by the lowest increase (0.04), and these average 0.075. This type of fluctuation may result from economic conditions;

   b. The annual increase averages 0.054 and the individual years are generally close to this average; therefore, a straight-line relationship is suggested;

   c. Comparing the last three increases (average 0.057) with the first three (average 0.05) reveals a slightly higher rate of growth that deviates from the straight-line concept;

   d. The percentage growth fluctuates considerably but appears to have a declining trend. This confirms that the exponential growth concept is, at best, small;

   e. The annual increase for toll traffic is not constant. The fact that it tends to be larger each year suggests an exponential growth pattern;

   f. The percent growth toll column indicates fluctuations, but tends to confirm the exponential nature of the growth. The average is 0.089. The second, third, and fourth year figures average 0.089 and the last three average 0.089. This suggests a relatively consistent exponential growth rate of about 8.9% per year.

Table 7.1 Ten-Year Originating and Toll Traffic Data

Year	Originating CCS/MS	Annual Increase	Percent Growth	Toll CCS/MS	Annual Increase	Percent Growth
1	1.60	—	—	0.150	—	—
2	1.65	0.05	0.031	0.179	0.029	0.193
3	1.70	0.05	0.030	0.195	0.016	0.089
4	1.75	0.05	0.029	0.212	0.017	0.087
5	1.86	0.11	0.063	0.231	0.019	0.090
6	1.90	0.04	0.022	0.252	0.021	0.091
7	1.97	0.07	0.037	0.275	0.023	0.091
8	2.03	0.06	0.032	0.299	0.024	0.087
9	2.08	0.05	0.025	0.326	0.026	0.090
10	2.14	0.06	0.029	0.356	0.030	0.092

The simplest technique for the evaluation of historical data is called *trending*; however, it is only applicable for straight-line growth. This method determines the rate of growth by comparing two or more values and the time span between them. The values used obviously can affect the results. The following examples use the data of Table 7.1 for originating traffic.

Comparing the first and last items, the calculation is

$$(2.14 - 1.60)/9 = 0.06 \text{ CCS/MS}$$

Comparing the first two and last two items, the results are

$$[(2.08 + 2.14) - (1.60 + 1.65)]/(2)(8) = 0.06 \text{ CCS/MS}$$

Comparing the first three and the last three items, the results are

$$[(2.03 + 2.08 + 2.14) - (1.60 + 1.65 + 1.70)]/(3)(7) = 0.062 \text{ CCS/MS}$$

Comparing years 5, 6, and 7 with 8, 9, and 10 results in

$$[(2.03 + 2.08 + 2.14) - (1.86 + 1.90 + 1.97)]/(3)(3) = 0.058 \text{ CCS/MS}$$

This technique suggests a relatively constant annual increase for the originating traffic (straight-line trend) that may be added to the base-year data using

$$y = a + bx \tag{7.1}$$

where

a = base-year data

b = annual increase

x = future year (following base year)

Another technique for the evaluation of historical data is called the *time-series method*. It involves the determination of (7.1) as illustrated in the following analysis of the originating traffic of Table 7.1, where (7.2) and (7.3) must be solved simultaneously (see Table 7.2). The years have been assumed as half years centered around five and a half years to force Σx to equal zero (necessary when there are an even number of samples).

Table 7.2 Linear Calculation Using Time-Series Method

x	y	xy	x^2
-9	1.60	-14.40	81
-7	1.65	-11.55	49
-5	1.70	-8.50	25
-3	1.75	-5.25	9
-1	1.86	-1.86	1
+1	1.90	+1.90	1
+3	1.97	+5.91	9
+5	2.03	+10.15	25
+7	2.08	+14.56	49
+9	2.14	+19.26	81
$\Sigma x = 0$	$\Sigma y = 18.68$	$\Sigma xy = 10.22$	$\Sigma x^2 = 330$

$$\Sigma y = Na + b\Sigma x \qquad (7.2)$$
$$\Sigma xy = a\Sigma x + b\Sigma x^2 \qquad (7.3)$$

$\Sigma y = Na + b\Sigma x$ $\Sigma xy = a\Sigma x + b\Sigma x^2$

$18.68 = 10a + 0$ $10.22 = 0 + 330b$

$a = 18.68/10 = 1.868$ $b = 10.22/330 = 0.031$

Using (7.1):

$$y = 1.868 + 0.031\, x \qquad (\text{base year} = 5\tfrac{1}{2})$$

To change the year "$5\tfrac{1}{2}$" to the most recent value (10), change a to 2.14, the value for the tenth year, and double the value of b (for full-year increments):

$$y = 2.14 + 0.062\,x \qquad (\text{base year} = 10)$$

When an exponential growth rate applies, an equation for the curve may be calculated by converting the values to logarithms and solving with the time-series method. Equation (7.4) represents the standard form of this equation. In logarithms, this converts to (7.5) which has a linear form, similar to that of (7.1):

$$z = cd^x \qquad\qquad\qquad (7.4)$$

$$\log z = \log c + (\log d)x \qquad\qquad (7.5)$$

This may be illustrated in the following analysis of the toll traffic data of Table 7.1. Equations (7.6) and (7.7) are the two equations that must be solved simultaneously, as with (7.2) and (7.3). The calculations are presented in Table 7.3, where the first-year is omitted because it is inconsistent with that of the other years.

Table 7.3 Exponential Calculation Using Time-Series Method

x	z	log z	x log z	x^2
-4	0.179	-0.7471	+2.9884	16
-3	0.195	-0.7100	+2.1300	9
-2	0.212	-0.6737	+1.3474	4
-1	0.231	-0.6364	+0.6364	1
0	0.252	-0.5986	0	0
+1	0.275	-0.5607	-0.5907	1
+2	0.299	-0.5243	-1.0486	4
+3	0.325	-0.4868	-1.4604	9
+4	0.356	-0.4486	-1.7944	6
$\Sigma x = 0$	—	$\Sigma \log z = -5.4162$	$\Sigma(x \log z) = +2.2081$	$\Sigma x^2 = 60$

$$\Sigma \log z = N \log c + (\log d)\Sigma x \qquad\qquad (7.6)$$

$$\Sigma(x \log z) = (\log c)\Sigma x + (\log d)\Sigma x^2 \qquad\qquad (7.7)$$

$$\Sigma \log z = N \log c + (\log d)\Sigma x \qquad \Sigma(x \log z) = (\log c)\Sigma x + (\log d)\Sigma x^2$$

$$-5.4162 = 9 \log c + 0 \qquad\qquad 2.2081 = 0 + 60 \log d$$

$$\log c = -5.4162/9 = -0.6018 \qquad \log d = 2.2081/60 = 0.0368$$

Using (7.5):

$$\log z = -0.6018 + 0.0368x \qquad \text{(base year = 6)}$$
$$\log z = -0.4486 + 0.0368x \qquad \text{(base year = 10)}$$

Using (7.4):

$$z = (0.250)(1.088)^x \qquad \text{(base year = 6)}$$
$$z = (0.356)(1.088)^x \qquad \text{(base year = 10)}$$

Note that the value of d (1.088) is not doubled when converting to year 10 (calculation based on full years rather than half years).

To determine if the *basic traffic data* is linear or exponential requires a calculation of both formulas and examination of the deviation from actual data to select the formula that best fits the data. Calculation of the *originating traffic data* using the exponential technique results in the following formula for base year 10:

$$y = (2.14)(1.034)^x \qquad \text{(base year = 10)}$$

The calculation of the deviation is given in Table 7.4 for both the linear and exponential traffic growth assumptions. The deviation for the linear equation is less than the deviation for the exponential equation; therefore, the linear equation fits the data better and should be used for projections.

Table 7.4 Originating Traffic Data Deviation

		Linear		Exponential	
Year	y	y'	(y-y')²	y'	(y-y')²
1	1.60	1.58	0.0004	1.58	0.0004
2	1.65	1.64	0.0001	1.64	0.0001
3	1.70	1.71	0.0001	1.69	0.0001
4	1.75	1.77	0.0004	1.75	0.0000
5	1.86	1.83	0.0009	1.81	0.0025
6	1.90	1.89	0.0001	1.87	0.0009
7	1.97	1.95	0.0004	1.94	0.0009
8	2.03	2.02	0.0001	2.00	0.0009
9	2.08	2.08	0.0000	2.07	0.0001
10	2.14	2.14	0.0000	2.14	0.0000
			$\Sigma(y-y')^2 = 0.0025$		$\Sigma(y-y')^2 = 0.0059$

where

y = Originating traffic from Table 7.1

y' = Originating traffic calculated from formulas

The same procedure may be followed for the *toll traffic data*. The results presented in Table 7.5 prove that the exponential equation fits the data better and should be used for projections.

Table 7.5 Toll Traffic Data Deviation

Year	y	Linear		Exponential	
		y'	(y-y')²	y'	(y-y')²
2	0.179	0.181	0.0004	0.181	0.0004
3	0.195	0.203	0.0064	0.197	0.0004
4	0.212	0.225	0.0169	0.214	0.0004
5	0.231	0.247	0.0256	0.233	0.0004
6	0.252	0.268	0.0256	0.254	0.0004
7	0.275	0.290	0.0225	0.276	0.0001
8	0.299	0.312	0.0169	0.301	0.0004
9	0.326	0.334	0.0064	0.327	0.0001
10	0.356	0.356	0.0000	0.356	0.0000
			$\Sigma(y-y')^2 = 0.1207$		$\Sigma(y-y')^2 = 0.0026$

where

y = Toll traffic from Table 7.1

y' = Toll traffic calculated from formulas

7.3.2 Estimating Future Trends

Although the future is likely to resemble the past in many respects, there may be predictable deviations that will require adjustment of the established trends. The equations derived from the preceding trend analyses generally are the starting point in forecasting. If no deviations can be foreseen, or if a deviation can be foreseen but its effect is in doubt, the forecast may be made using the past trend without adjustment. However, this decision should be noted so that the forecast may be revised when experience or further study provides a better basis for adjustment.

New services or features lead to altered calling habits causing traffic characteristics and growth rates to change. Frequently, the rate of growth is diminished (tapered) by the forecaster because of the personal opinion that past rates of growth cannot be sustained indefinitely. While it may be true that a saturation point will be reached, there is little such indication today. The forecaster cannot be prohibited from such opinion, but should always prepare an untapered forecast as an upper limit for planning. Any plan that suffices for both tapered and untapered forecasts has flexibility for growth.

The trend formula must be adjusted when the rate of growth is expected to change. Traffic volume must be adjusted if a discontinuity is expected with the introduction of a new service or feature—the volume may be increased or decreased accordingly. If this also alters the growth rate, the trend formula must be readjusted. In some cases, the calling rate may need to be adjusted rather than the traffic volume.

The adjustments may be made in the base-year data or in the fore-casted data. It is generally more convenient to adjust the base-year data and use a single formula to project to all future years. Where contingency forecasts are to be made, the following techniques are suggested:

a. For the assumption than an event will not occur, there may be no need to adjust either base-year or the growth rate;
b. For the assumption that an event will occur, the base-year data and, if necessary, the trend formula should be adjusted and a new pro-jection made;
c. The above projections result in two separate curves for the entire forecast period, which may be directly compared to identify the effect of the contingency in some future year. This also permits studies to determine the optimum timing, or the penalty incurred for incorrect timing.

Occasionally, line and main-station forecasts are questionable. These forecasts are of extreme importance in all planning activities, and the only realistic answer is to have them revised rather than attempt to circumvent them. Such forecasts must determine the number of terminations that are to be provided in an office. It would be ridiculous to provide sufficient traffic capacity without adequate termination capacity.

In summary, formulas of growth trends incorporate the base-year volume as one of the years. Solving the trend formulas for future years gives the projected values for those years in terms of the parameters being projected. If base-year adjustments are required, the formulas must be modified accordingly. Computer programs have been written for process-ing these data and developing the engineering base. This should be the starting point for all forecasting. Naturally, it should be compared with other historical data before being used to guard against errors or abnormal conditions. The normal procedure is to plot the results on the historical curves for visual evaluation.

7.3.3 Professional Judgment

In most systems, tables and graphs have been used to determine requirements based on future estimated loads, calls, or main stations. However, service criteria may not be met even with the best estimates if judgment is not used when applying the published material. Professional judgment depends on an understanding of the basis for the data, and how, when, and where it should be used.

When individual circumstances differ from system characteristics, we have the responsibility for adjusting quantities and arrangements to meet these circumstances. Furthermore, we must be able to support those adjustments to management, that is, a careful analysis of the data and the system must be made to justify any deviation from standards. If there is a problem based on table assumptions, the analysis will aid in pointing to the logical reasons why the table doesn't fit and, at least qualitatively, what further adjustments should be made.

In any case where judgment indicates that an empirical result should be used instead of published standards, it is important that such empirical data include results well into the overloaded area, and that the data are in sufficient quantity to be representative. Where adequate overload data are not available, careful predictions must be made as to what is likely to happen when such overloads do occur.

Appendix A
Poisson Traffic Capacity Tables

The Poisson traffic capacity tables are used in North America to dimension final trunk groups (see Section 3.3). The Poisson distribution is based on the following assumptions:

a. infinite sources;
b. blocked calls held;
c. constant or exponential holding time;
d. full availability.

In the tables, P is used to represent the blocking probability, A is used to represent traffic capacity of the group in Erlangs, and N is used to represent the number of trunks in the group.

Poisson Traffic Capacities

N				A for P =				N
	.001	.002	.005	.010	.020	.050	.100	
1	.001	.002	.005	.011	.021	.053	.106	1
2	.044	.065	.104	.150	.214	.358	.531	2
3	.192	.244	.338	.436	.567	.817	1.10	3
4	.428	.519	.673	.822	1.02	1.37	1.75	4
5	.739	.868	1.08	1.28	1.55	1.97	2.44	5
6	1.11	1.27	1.54	1.79	2.11	2.61	3.14	6
7	1.52	1.72	2.04	2.33	2.69	3.28	3.89	7
8	1.97	2.21	2.57	2.91	3.31	3.97	4.67	8
9	2.45	2.72	3.13	3.50	3.94	4.69	5.42	9
10	2.97	3.26	3.72	4.14	4.61	5.42	6.22	10
11	3.50	3.82	4.32	4.78	5.31	6.17	7.03	11
12	4.03	4.40	4.94	5.43	6.00	6.92	7.83	12
13	4.61	5.00	5.58	6.11	6.69	7.69	8.64	13
14	5.19	5.61	6.23	6.78	7.42	8.47	9.47	14
15	5.78	6.23	6.89	7.47	8.14	9.25	10.3	15
16	6.42	6.87	7.57	8.18	8.89	10.1	11.1	16
17	7.03	7.52	8.25	8.89	9.64	10.8	12.0	17
18	7.67	8.17	8.94	9.61	10.4	11.6	12.8	18
19	8.31	8.84	9.65	10.4	11.1	12.4	13.7	19
20	8.97	9.52	10.4	11.1	11.9	13.3	14.5	20
21	9.61	10.2	11.1	11.8	12.7	14.1	15.4	21
22	10.3	10.9	11.8	12.6	13.5	14.9	16.3	22
23	11.0	11.6	12.5	13.3	14.3	15.7	17.1	23
24	11.6	12.3	13.3	14.1	15.1	16.6	18.0	24
25	12.3	13.0	14.0	14.9	15.9	17.4	18.8	25
26	13.0	13.7	14.7	15.6	16.6	18.2	19.7	26
27	13.8	14.4	15.5	16.4	17.4	19.1	20.6	27
28	14.4	15.2	16.3	17.2	18.2	19.9	21.5	28
29	15.1	15.9	17.0	18.0	19.0	20.8	22.4	29
30	15.9	16.6	17.8	18.8	19.9	21.6	23.2	30
31	16.6	17.4	18.5	19.5	20.7	22.5	24.1	31
32	17.3	18.1	19.3	20.3	21.5	23.3	25.0	32
33	18.1	18.9	20.1	21.1	22.3	24.2	25.9	33
34	18.8	19.6	20.9	21.9	23.1	25.1	26.8	34
35	19.5	20.4	21.6	22.7	23.9	25.9	27.7	35
36	20.3	21.1	22.4	23.5	24.8	26.8	28.6	36
37	21.0	21.9	23.2	24.3	25.6	27.6	29.4	37
38	21.8	22.7	24.0	25.1	26.4	28.5	30.3	38
39	22.5	23.4	24.8	26.0	27.3	29.4	31.3	39
40	23.3	24.2	25.6	26.8	28.1	30.2	32.1	40

Poisson Traffic Capacities (Continued)

N	.001	.002	.005	.010	.020	.050	.100	N
				A for P =				
41	24.0	25.0	26.4	27.6	28.9	31.1	33.1	41
42	24.8	25.8	27.2	28.4	29.8	32.0	33.9	42
43	25.5	26.5	28.0	29.2	30.6	32.9	34.9	43
44	26.3	27.3	28.8	30.1	31.5	33.7	35.8	44
45	27.1	28.1	29.6	30.9	32.3	34.6	36.7	45
46	27.9	28.9	30.4	31.7	33.2	35.5	37.6	46
47	28.6	29.7	31.2	32.5	34.0	36.4	38.5	47
48	29.4	30.5	32.0	33.4	34.9	37.2	39.4	48
49	30.2	31.3	32.9	34.2	35.7	38.1	40.3	49
50	31.0	32.1	33.7	35.0	36.6	39.0	41.2	50
51	31.8	32.9	34.5	35.9	37.4	39.9	42.1	51
52	32.5	33.7	35.3	36.7	38.3	40.8	43.0	52
53	33.3	34.5	36.1	37.6	39.2	41.6	43.9	53
54	34.1	35.3	37.0	38.4	40.0	42.5	44.8	54
55	34.9	36.1	37.8	39.2	40.9	43.4	45.7	55
56	35.7	36.9	38.6	40.1	41.8	44.3	46.6	56
57	36.5	37.7	39.3	40.9	42.6	45.2	47.6	57
58	37.3	38.5	40.3	41.8	43.5	46.1	48.5	58
59	38.1	39.3	41.1	42.6	44.3	47.0	49.4	59
60	38.9	40.1	41.9	43.5	45.2	47.9	50.3	60
61	39.7	40.9	42.8	44.3	46.1	48.8	51.2	61
62	40.5	41.8	43.6	45.2	46.9	49.6	52.1	62
63	41.3	42.6	44.4	46.0	47.8	50.5	53.1	63
64	42.1	43.4	45.3	46.9	48.7	51.4	54.0	64
65	42.9	44.2	46.1	47.7	49.6	52.3	54.9	65
66	43.7	45.0	47.9	48.6	50.4	53.2	55.8	66
67	44.5	45.9	47.8	49.4	51.3	54.1	56.7	67
68	45.3	46.7	48.7	50.3	52.2	55.0	57.7	68
69	46.1	47.5	49.5	51.2	53.1	55.9	58.6	69
70	47.0	48.4	50.3	52.0	53.9	56.8	59.5	70
71	47.8	49.2	51.2	52.9	54.8	57.7	60.4	71
72	48.6	50.0	52.0	53.8	55.7	58.6	61.4	72
73	49.4	50.8	52.9	54.6	56.6	59.5	62.3	73
74	50.3	51.7	53.7	55.5	57.4	60.4	63.2	74
75	51.1	52.5	54.6	56.3	58.3	61.3	64.1	75
76	51.9	53.4	55.4	57.2	59.2	62.3	65.1	76
77	52.7	54.2	56.3	58.1	60.1	63.2	66.0	77
78	53.5	55.0	57.1	58.9	60.9	64.1	66.9	78
79	54.4	55.9	58.0	59.8	61.8	65.0	67.9	79
80	55.2	56.7	58.9	60.7	62.7	65.9	68.9	80

Poisson Traffic Capacities (Continued)

N	A for P =							N
	.001	.002	.005	.010	.020	.050	.100	
81	56.0	57.5	59.7	61.5	63.6	66.8	69.7	81
82	56.8	58.4	60.6	62.4	64.5	67.7	70.6	82
83	57.7	59.2	61.4	63.3	65.4	68.6	71.6	83
84	58.5	60.1	62.3	64.2	66.3	69.5	72.5	84
85	59.3	60.9	63.1	65.0	67.2	70.4	73.4	85
86	60.2	61.8	64.0	65.9	68.1	71.4	74.4	86
87	61.0	62.6	64.9	66.8	68.9	72.3	75.3	87
88	61.8	63.5	65.7	67.7	69.8	73.2	76.3	88
89	62.7	64.3	66.6	68.5	70.7	74.1	77.2	89
90	63.5	65.2	67.5	69.4	71.6	75.0	78.1	90
91	64.4	66.0	68.3	70.3	72.5	75.9	79.1	91
92	65.2	66.9	69.2	71.2	73.4	76.8	80.0	92
93	66.0	67.7	70.2	72.1	74.3	77.7	80.9	93
94	66.9	68.6	70.9	72.9	75.2	78.6	81.9	94
95	67.7	69.4	71.8	73.8	76.1	79.6	82.8	95
96	68.6	70.3	72.7	74.7	77.0	80.5	83.7	96
97	69.4	71.1	73.5	75.6	77.9	81.4	84.7	97
98	70.2	72.0	74.4	76.4	78.8	82.3	85.6	98
99	71.1	72.8	75.3	77.3	79.7	83.2	86.6	99
100	71.9	73.7	76.2	78.2	80.6	84.1	87.5	100
101	72.8	74.6	77.0	79.1	81.4	85.1	88.3	101
102	73.6	75.4	77.9	80.0	82.4	86.0	89.3	102
103	74.5	76.3	78.8	80.8	83.2	86.9	90.2	103
104	75.3	77.1	79.6	81.7	84.1	87.8	91.2	104
105	76.2	78.0	80.5	82.6	85.0	88.7	92.1	105
106	77.0	78.9	81.4	83.5	85.9	89.6	93.0	106
107	77.9	79.7	82.3	84.4	86.8	90.6	94.0	107
108	78.7	80.6	83.2	85.3	87.7	91.5	94.9	108
109	79.6	81.4	84.0	86.2	88.6	92.4	95.8	109
110	80.4	82.3	84.9	87.1	89.5	93.3	96.8	110
111	81.3	83.2	85.8	87.9	90.4	94.3	97.7	111
112	82.1	84.0	86.7	88.8	91.3	95.2	98.7	112
113	83.0	84.9	87.5	89.7	92.3	96.1	99.6	113
114	83.8	85.8	88.4	90.6	93.1	97.0	100.5	114
115	84.7	86.6	89.3	91.5	94.1	97.9	101.5	115
116	85.6	87.5	90.2	92.4	94.9	98.9	102.4	116
117	86.4	88.4	91.1	93.3	95.9	99.8	103.4	117
118	87.3	89.3	92.0	94.2	96.8	100.7	104.3	118
119	88.1	90.1	92.8	95.1	97.7	101.6	105.3	119
120	89.0	91.0	93.7	96.0	98.6	102.6	106.2	120

Poisson TrafficCapacities (Continued)

N	.001	.002	.005	.010	.020	.050	.100	N
121	89.8	91.9	94.6	96.9	99.5	103.5	107.1	121
122	90.7	92.7	95.5	97.8	100.4	104.4	108.1	122
123	91.6	93.6	96.4	98.7	101.3	105.3	109.0	123
124	92.4	94.5	97.3	99.6	102.2	106.3	109.9	124
125	93.3	95.4	98.1	100.4	103.1	107.2	110.9	125
126	94.1	96.2	99.0	101.3	104.0	108.1	111.8	126
127	95.0	97.1	99.9	102.3	104.9	109.0	112.8	127
128	95.9	98.0	100.8	103.1	105.8	110.0	113.7	128
129	96.7	98.9	101.7	104.1	106.8	110.9	114.7	129
130	97.6	99.7	102.6	104.9	107.6	111.8	115.6	130
131	98.4	100.6	103.5	105.8	108.6	112.8	116.6	131
132	99.3	101.5	104.3	106.8	109.5	113.7	117.5	132
133	100.2	102.4	105.2	107.6	110.4	114.6	118.4	133
134	101.1	103.2	106.1	108.5	111.3	115.5	119.4	134
135	101.9	104.1	107.0	109.4	112.2	116.4	120.3	135
136	102.8	105.0	107.9	110.3	113.1	117.4	121.3	136
137	103.7	105.9	108.8	111.2	114.0	118.3	122.2	137
138	104.5	106.8	109.7	112.1	114.9	119.3	123.2	138
139	105.4	107.6	110.6	113.1	115.9	120.2	124.1	139
140	106.3	108.5	111.5	113.9	116.8	121.1	125.1	140
141	107.1	109.4	112.4	114.8	117.7	122.1	126.0	141
142	108.0	110.3	113.3	115.8	118.6	123.0	126.9	142
143	108.9	111.2	114.2	116.6	119.5	123.9	127.9	143
144	109.8	112.0	115.0	117.5	120.4	124.8	128.8	144
145	110.6	112.9	115.9	118.4	121.3	125.8	129.9	145
146	111.5	113.8	116.8	119.4	122.3	126.7	130.7	146
147	112.4	114.7	117.7	120.3	123.2	127.6	131.7	147
148	113.2	115.6	118.6	121.2	124.1	128.6	132.6	148
149	114.1	116.5	119.5	122.1	125.0	129.5	133.6	149
150	114.9	117.4	120.4	123.0	125.9	130.4	134.5	150
151	115.9	118.2	121.3	123.9	126.8	131.4	135.5	151
152	116.7	119.1	122.2	124.8	127.8	132.3	136.4	152
153	117.6	120.0	123.1	125.7	128.7	133.2	137.4	153
154	118.5	120.9	124.0	126.6	129.6	134.2	138.3	154
155	119.4	121.8	124.9	127.5	130.5	135.1	139.3	155
156	120.2	122.7	125.8	128.4	131.4	136.0	140.2	156
157	121.1	123.6	126.7	129.3	132.3	137.0	141.2	157
158	122.0	124.5	127.6	130.2	133.3	137.9	142.1	158
159	122.9	125.3	128.5	131.1	134.2	138.8	143.1	159
160	123.8	126.2	129.4	132.1	135.1	139.8	144.0	160

Appendix B
Erlang-B Traffic Capacity Tables

The Erlang-B traffic capacity tables are used in North America to dimension high-usage trunk groups and internationally to dimension all trunk groups (see Section 3.4). The Erlang-B distribution is based on the following assumptions:

a. infinite sources;
b. blocked calls cleared;
c. constant or exponential holding time;
d. full availability.

In the tables, P is used to represent the blocking probability, A is used to represent traffic capacity of the group in Erlangs, and N is used to represent the number of trunks in the group.

Erlang-B Traffic Capacities

N	.001	.002	.005	.010	.020	.050	.100	N
				A for P =				
1	.001	.002	.005	.011	.021	.053	.111	1
2	.046	.066	.106	.153	.224	.382	.595	2
3	.194	.249	.349	.456	.603	.900	1.27	3
4	.440	.536	.702	.870	1.09	1.52	2.05	4
5	.763	.900	1.13	1.36	1.66	2.22	2.88	5
6	1.15	1.33	1.62	1.91	2.28	2.96	3.76	6
7	1.58	1.80	2.16	2.50	2.94	3.74	4.67	7
8	2.05	2.31	2.73	3.13	3.63	4.54	5.60	8
9	2.56	2.86	3.33	3.78	4.34	5.37	6.55	9
10	3.09	3.43	3.96	4.46	5.08	6.22	7.51	10
11	3.65	4.02	4.61	5.16	5.84	7.08	8.49	11
12	4.23	4.64	5.28	5.88	6.62	7.95	9.47	12
13	4.83	5.27	5.96	6.61	7.41	8.83	10.5	13
14	5.45	5.92	6.66	7.35	8.20	9.73	11.5	14
15	6.08	6.58	7.38	8.11	9.01	10.6	12.5	15
16	6.72	7.26	8.10	8.87	9.83	11.5	13.5	16
17	7.38	7.95	8.83	9.65	10.7	12.5	14.5	17
18	8.05	8.64	9.58	10.4	11.5	13.4	15.6	18
19	8.72	9.35	10.3	11.2	12.3	14.3	16.6	19
20	9.41	10.1	11.1	12.0	13.2	15.3	17.6	20
21	10.1	10.8	11.9	12.8	14.0	16.2	18.7	21
22	10.8	11.5	12.6	13.7	14.9	17.1	19.7	22
23	11.5	12.3	13.4	14.5	15.8	18.1	20.7	23
24	12.2	13.0	14.2	15.3	16.6	19.0	21.8	24
25	13.0	13.8	15.0	16.1	17.5	20.0	22.8	25
26	13.7	14.5	15.8	17.0	18.4	20.9	23.9	26
27	14.4	15.3	16.6	17.8	19.3	21.9	24.9	27
28	15.2	16.1	17.4	18.6	20.2	22.9	26.0	28
29	15.9	16.8	18.2	19.5	21.0	23.8	27.1	29
30	16.7	17.6	19.0	20.3	21.9	24.8	28.1	30
31	17.4	18.4	19.9	21.2	22.8	25.8	29.2	31
32	18.2	19.2	20.7	22.1	23.7	26.8	30.2	32
33	19.0	20.0	21.5	22.9	24.6	27.7	31.3	33
34	19.7	20.8	22.3	23.8	25.5	28.7	32.4	34
35	20.5	21.6	23.2	24.6	26.4	29.7	33.4	35
36	21.3	22.4	24.0	25.5	27.3	30.7	34.5	36
37	22.1	23.2	24.9	26.4	28.3	31.6	35.6	37
38	22.9	24.0	25.7	27.3	29.2	32.6	36.6	38
39	23.7	24.8	26.5	28.1	30.1	33.6	37.7	39
40	24.4	25.6	27.4	29.0	31.0	34.6	38.8	40

Erlang-B Traffic Capacities

N	.001	.002	.005	.010	.020	.050	100	N
				A for P =				
41	25.2	26.4	28.2	29.9	31.9	35.6	39.9	41
42	26.0	27.2	29.1	30.8	32.8	36.6	40.9	42
43	26.8	28.1	29.9	31.7	33.8	37.6	42.0	43
44	27.6	28.9	30.8	32.5	34.7	38.6	43.1	44
45	28.5	29.7	31.7	33.4	35.6	39.6	44.2	45
46	29.3	30.5	32.5	34.3	36.5	40.5	45.2	46
47	30.1	31.4	33.4	35.2	37.5	41.5	46.3	47
48	30.9	32.2	34.2	36.1	38.4	42.5	47.4	48
49	31.7	33.0	35.1	37.0	39.3	43.5	48.5	49
50	32.5	33.9	36.0	37.9	40.3	44.5	49.6	50
51	33.3	34.7	36.9	38.8	41.2	45.5	50.6	51
52	34.2	35.6	37.7	39.7	42.1	46.5	51.7	52
53	35.0	36.4	38.6	40.6	43.1	47.5	52.8	53
54	35.8	37.3	39.5	41.5	44.0	48.5	53.9	54
55	36.6	38.1	40.4	42.4	44.9	49.5	55.0	55
56	37.5	38.9	41.2	43.3	45.9	50.5	56.1	56
57	38.3	39.8	42.1	44.2	46.8	51.5	57.1	57
58	39.1	40.6	43.0	45.1	47.8	52.6	58.2	58
59	40.0	41.5	43.9	46.0	48.7	53.6	59.3	59
60	40.8	42.4	44.8	46.9	49.6	54.6	60.4	60
61	41.6	43.2	45.6	47.9	50.6	55.6	61.5	61
62	42.5	44.1	46.5	48.8	51.5	56.6	62.6	62
63	43.3	44.9	47.7	49.7	52.5	57.6	63.7	63
64	44.2	45.8	48.3	50.6	53.4	58.6	64.8	64
65	45.0	46.7	49.2	51.5	54.4	59.6	65.8	65
66	45.8	47.5	50.1	52.4	55.3	60.6	66.9	66
67	46.7	48.4	51.0	53.4	56.3	61.6	68.0	67
68	47.5	49.2	51.9	54.3	57.2	62.6	69.1	68
69	48.4	50.1	52.8	55.2	58.2	63.7	70.2	69
70	49.2	51.0	53.7	56.1	59.1	64.7	71.3	70
71	50.1	51.9	54.6	57.0	60.1	65.7	72.4	71
72	50.9	52.7	55.5	58.0	61.0	66.7	73.5	72
73	51.8	53.6	56.4	58.9	62.0	67.7	74.6	73
74	52.7	54.5	57.3	59.8	62.9	68.7	75.6	74
75	53.5	55.3	58.2	60.7	63.9	69.7	76.7	75
76	54.4	56.2	59.1	61.7	64.9	70.8	77.8	76
77	55.2	57.1	60.0	62.6	65.8	71.8	78.9	77
78	56.1	58.0	60.9	63.5	66.8	72.8	80.0	78
79	57.0	58.8	61.8	64.4	67.7	73.8	81.1	79
80	57.8	59.7	62.7	65.4	68.7	74.8	82.2	80

Erlang-B Traffic Capacities

	A for P =							
N	.001	.002	.005	.010	.020	.050	.100	N
81	58.7	60.6	63.6	66.3	69.6	75.8	83.3	81
82	59.5	61.5	64.5	67.2	70.6	76.9	84.4	82
83	60.4	62.4	65.4	68.2	71.6	77.9	85.5	83
84	61.3	63.2	66.3	69.1	72.5	78.9	86.6	84
85	62.1	64.1	67.2	70.0	73.5	79.9	87.7	85
86	63.0	65.0	68.1	70.9	74.5	80.9	88.8	86
87	63.9	65.9	69.0	71.9	75.4	82.0	89.9	87
88	64.7	66.8	69.9	72.8	76.4	83.0	91.0	88
89	65.6	67.7	70.8	73.7	77.3	84.0	92.1	89
90	66.5	68.6	71.8	74.7	78.3	85.0	93.1	90
91	67.4	69.4	72.7	75.6	79.3	86.0	94.2	91
92	68.2	70.3	73.6	76.6	80.2	87.1	95.3	92
93	69.1	71.2	74.5	77.5	81.2	88.1	96.4	93
94	70.0	72.1	75.4	78.4	82.2	89.1	97.5	94
95	70.9	73.0	76.3	79.4	83.1	90.1	98.6	95
96	71.7	73.9	77.2	80.3	84.1	91.1	99.7	96
97	72.6	74.8	78.2	81.2	85.1	92.2	100.8	97
98	73.5	75.7	79.1	82.2	86.0	93.2	101.9	98
99	74.4	76.6	80.0	83.1	87.0	94.2	103.0	99
100	75.2	77.5	80.9	84.1	88.0	95.2	104.1	100
101	76.1	78.4	81.8	85.0	88.9	96.3	105.2	101
102	77.0	79.3	82.8	85.9	89.9	97.3	106.3	102
103	77.9	80.2	83.7	86.9	90.9	98.3	107.4	103
104	78.8	81.1	84.6	87.8	91.9	99.3	108.5	104
105	79.7	81.9	85.5	88.8	92.8	100.4	109.6	105
106	80.5	82.9	86.4	89.7	93.8	101.4	110.7	106
107	81.4	83.8	87.4	90.7	94.8	102.4	111.8	107
108	82.3	84.7	88.3	91.6	95.7	103.4	112.9	108
109	83.2	85.6	89.2	92.5	96.7	104.5	114.0	109
110	84.1	86.5	90.1	93.5	97.7	105.5	115.1	110
111	85.0	87.4	91.0	94.4	98.7	106.5	116.2	111
112	85.9	88.3	92.0	95.4	99.6	107.5	117.3	112
113	86.7	89.2	92.9	96.3	100.6	108.6	118.4	113
114	87.6	90.1	93.8	97.3	101.6	109.6	119.5	114
115	88.5	91.0	94.7	98.2	102.5	110.6	120.6	115
116	89.4	91.9	95.7	99.2	103.5	111.7	121.7	116
117	90.3	92.8	96.6	100.1	104.5	112.7	122.8	117
118	91.2	93.7	97.5	101.1	105.5	113.7	123.9	118
119	92.1	94.6	98.5	102.0	106.4	114.7	125.0	119
120	93.0	95.5	99.4	103.0	107.4	115.8	126.1	120

Erlang-B Traffic Capacities

N	A for P =							N
	.001	.002	.005	.010	.020	.050	.100	
121	93.8	96.4	100.3	103.9	108.4	116.8	127.2	121
122	94.7	97.3	101.2	104.9	109.4	117.8	128.3	122
123	95.6	98.2	102.2	105.8	110.3	118.9	129.4	123
124	96.5	99.1	103.1	106.8	111.3	119.9	130.5	124
125	97.4	100.0	104.0	107.7	112.3	120.9	131.6	125
126	98.3	100.9	105.0	108.7	113.3	122.0	132.7	126
127	99.2	101.8	105.9	109.6	114.2	123.0	133.8	127
128	100.1	102.7	106.8	110.6	115.2	124.0	134.9	128
129	101.0	103.7	107.8	111.5	116.2	125.1	136.0	129
130	101.9	104.6	108.7	112.5	117.2	126.1	137.1	130
131	102.8	105.5	109.6	113.4	118.2	127.1	138.2	131
132	103.7	106.4	110.5	114.4	119.1	128.2	139.3	132
133	104.6	107.3	111.5	115.3	120.1	129.2	140.4	133
134	105.5	108.2	112.4	116.3	121.1	130.2	141.5	134
135	106.4	109.1	113.3	117.2	122.1	131.3	142.6	135
136	107.3	110.0	114.3	118.2	123.1	132.3	143.7	136
137	108.2	111.0	115.2	119.1	124.0	133.3	144.8	137
138	109.1	111.9	116.2	120.1	125.0	134.3	145.9	138
139	110.0	112.8	117.1	121.0	126.0	135.4	147.0	139
140	110.9	113.7	118.0	122.0	127.0	136.4	148.1	140
141	111.8	114.6	119.0	122.9	128.0	137.4	149.2	141
142	112.7	115.5	119.9	123.9	128.9	138.5	150.3	142
143	113.6	116.4	120.8	124.8	129.9	139.5	151.4	143
144	114.5	117.4	121.8	125.8	130.9	140.5	152.5	144
145	115.4	118.3	122.7	126.7	131.9	141.6	153.6	145
146	116.3	119.2	123.6	127.7	132.9	142.6	154.7	146
147	117.2	120.1	124.6	128.6	133.8	143.6	155.8	147
148	118.1	121.0	125.5	129.6	134.8	144.7	156.9	148
149	119.0	121.9	126.5	130.6	135.8	145.7	158.0	149
150	119.9	122.9	127.4	131.6	136.8	146.7	159.1	150
151	120.8	123.8	128.3	132.5	137.8	147.8	160.3	151
152	121.7	124.7	129.3	133.5	138.7	148.8	161.4	152
153	122.6	125.6	130.2	134.5	139.7	149.8	162.5	153
154	123.5	126.5	131.2	135.4	140.7	150.9	163.6	154
155	124.4	127.5	132.1	136.4	141.7	150.9	164.7	155
156	125.3	128.4	133.0	137.4	142.7	152.9	165.8	156
157	126.2	129.3	134.0	138.3	143.6	153.9	166.9	157
158	127.1	130.2	134.9	139.3	144.6	154.9	168.0	158
159	128.0	131.1	135.9	140.3	145.6	156.0	169.1	159
160	129.0	132.1	136.8	141.2	146.6	157.0	170.2	160

Appendix C
Erlang-C Delay Loss Probability Tables

The Erlang-C delay loss probability tables are used to dimension common-equipment server pools in which calls that are not served immediately are held in queue until an idle server is available (see Section 3.5). The Erlang-C distribution is based on the following assumptions:

a. infinite sources;
b. blocked calls delayed;
c. exponential holding time;
d. full availability.

In the tables, $P(>0)$ is used to represent probability of delay greater than zero, $P(>t)$ is used to represent probability of delay greater than T_1, A is used to represent traffic offered to the group in Erlangs, N is used to represent the number of servers in the group, and T_1/T_2 is the delay-to-holding-time ratio.

Erlang-C Delay Loss Probabilities

(N = 2 Servers)

A	P(>0)	\(P(>t)\) for \(T_1/T_2 =\) 0.1	0.2	0.3	0.4	0.5	0.6	0.7	0.8	0.9	1.0	A
0.02	.0002	.0002	.0001	.0001	.0001	.0001	.0001					0.02
0.04	.0008	.0006	.0005	.0004	.0004	.0003	.0002	.0002	.0002	.0001	.0001	0.04
0.06	.0017	.0014	.0012	.0010	.0008	.0007	.0005	.0004	.0004	.0003	.0003	0.06
0.08	.0031	.0025	.0021	.0017	.0014	.0012	.0010	.0008	.0007	.0005	.0005	0.08
0.10	.0048	.0039	.0033	.0027	.0022	.0018	.0015	.0013	.0010	.0009	.0007	0.10
0.12	.0068	.0056	.0047	.0039	.0032	.0027	.0022	.0018	.0015	.0013	.0010	0.12
0.14	.0092	.0076	.0063	.0052	.0044	.0036	.0030	.0025	.0021	.0017	.0014	0.14
0.16	.0119	.0099	.0082	.0068	.0057	.0047	.0039	.0033	.0027	.0023	.0019	0.16
0.18	.0149	.0124	.0103	.0086	.0072	.0060	.0050	.0042	.0035	.0029	.0024	0.18
0.20	.0182	.0152	.0127	.0106	.0089	.0074	.0062	.0052	.0043	.0036	.0030	0.20
0.22	.0218	.0182	.0153	.0128	.0107	.0090	.0075	.0063	.0052	.0044	.0037	0.22
0.24	.0257	.0216	.0181	.0152	.0127	.0107	.0089	.0075	.0063	.0053	.0044	0.24
0.26	.0299	.0251	.0211	.0177	.0149	.0125	.0105	.0088	.0074	.0062	.0053	0.26
0.28	.0344	.0290	.0244	.0205	.0173	.0146	.0123	.0103	.0087	.0073	.0062	0.28
0.30	.0391	.0330	.0279	.0235	.0198	.0167	.0141	.0119	.0100	.0085	.0071	0.30
0.32	.0441	.0373	.0315	.0267	.0225	.0191	.0161	.0136	.0115	.0097	.0082	0.32
0.34	.0494	.0418	.0354	.0300	.0254	.0215	.0182	.0155	.0131	.0111	.0094	0.34
0.36	.0549	.0466	.0396	.0336	.0285	.0242	.0205	.0174	.0148	.0126	.0107	0.36
0.38	.0607	.0515	.0439	.0373	.0320	.0270	.0229	.0196	.0166	.0141	.0120	0.38
0.40	.0667	.0568	.0484	.0413	.0352	.0300	.0255	.0218	.0185	.0158	.0135	0.40
0.42	.0729	.0622	.0531	.0454	.0387	.0331	.0282	.0241	.0206	.0176	.0150	0.42
0.44	.0793	.0679	.0581	.0497	.0425	.0364	.0311	.0266	.0228	.0195	.0167	0.44
0.46	.0860	.0737	.0632	.0542	.0465	.0398	.0341	.0293	.0251	.0215	.0184	0.46
0.48	.0929	.0798	.0685	.0589	.0506	.0434	.0373	.0321	.0275	.0237	.0203	0.48
0.50	.1000	.0861	.0741	.0638	.0549	.0472	.0407	.0350	.0301	.0259	.0223	0.50

Note: Blanks indicate \(P(>t) < 0.00005\)

Erlang-C Delay Loss Probabilities

(N = 3 Servers)

A	P(>0)	\multicolumn{11}{c}{P(>t) for T_1/T_2 =}										A
		0.1	0.2	0.3	0.4	0.5	0.6	0.7	0.8	0.9	1.0	
0.10	.0002	.0001	.0001	.0001								0.10
0.15	.0005	.0004	.0003	.0002	.0002	.0001	.0001	.0001	.0001			0.15
0.20	.0012	.0009	.0007	.0005	.0004	.0003	.0002	.0002	.0001	.0001	.0001	0.20
0.25	.0022	.0017	.0013	.0010	.0007	.0006	.0004	.0003	.0002	.0002	.0001	0.25
0.30	.0037	.0028	.0022	.0016	.0013	.0010	.0007	.0006	.0004	.0003	.0002	0.30
0.35	.0057	.0044	.0034	.0026	.0020	.0015	.0012	.0009	.0007	.0005	.0004	0.35
0.40	.0082	.0064	.0049	.0038	.0029	.0022	.0017	.0013	.0010	.0008	.0006	0.40
0.45	.0114	.0088	.0068	.0053	.0041	.0032	.0025	.0019	.0015	.0011	.0009	0.45
0.50	.0152	.0118	.0092	.0072	.0056	.0043	.0034	.0026	.0021	.0016	.0012	0.50
0.55	.0196	.0153	.0120	.0094	.0073	.0057	.0045	.0035	.0028	.0022	.0017	0.55
0.60	.0247	.0194	.0153	.0120	.0094	.0074	.0058	.0046	.0036	.0028	.0022	0.60
0.65	.0304	.0241	.0190	.0150	.0119	.0094	.0074	.0059	.0046	.0037	.0029	0.65
0.70	.0369	.0293	.0233	.0185	.0147	.0117	.0093	.0074	.0059	.0047	.0037	0.70
0.75	.0441	.0352	.0281	.0225	.0179	.0143	.0114	.0091	.0073	.0058	.0046	0.75
0.80	.0520	.0418	.0335	.0269	.0216	.0173	.0139	.0112	.0090	.0072	.0058	0.80
0.85	.0607	.0489	.0395	.0318	.0257	.0207	.0167	.0135	.0109	.0088	.0071	0.85
0.90	.0700	.0568	.0460	.0373	.0302	.0245	.0199	.0161	.0131	.0106	.0086	0.90
0.95	.0801	.0653	.0532	.0433	.0353	.0287	.0234	.0191	.0155	.0127	.0103	0.95
1.00	.0909	.0744	.0609	.0499	.0408	.0334	.0274	.0224	.0184	.0150	.0123	1.00
1.05	.1024	.0843	.0693	.0571	.0470	.0386	.0318	.0262	.0215	.0177	.0146	1.05
1.10	.1146	.0948	.0784	.0648	.0536	.0443	.0367	.0303	.0251	.0207	.0171	1.10
1.15	.1276	.1060	.0881	.0732	.0609	.0506	.0420	.0349	.0290	.0241	.0201	1.15
1.20	.1412	.1179	.0985	.0823	.0687	.0574	.0479	.0400	.0334	.0279	.0233	1.20
1.25	.1555	.1305	.1096	.0920	.0772	.0648	.0544	.0457	.0383	.0322	.0270	1.25
1.30	.1704	.1438	.1213	.1023	.0863	.0728	.0615	.0519	.0437	.0369	.0311	1.30

Note: Blanks indicate P(>t) < 0.00005

Erlang-C Delay Loss Probabilities

(N = 4 Servers)

A	P(>0)	P(>t) for T_1/T_2 =										A
		0.1	0.2	0.3	0.4	0.5	0.6	0.7	0.8	0.9	1.0	
0.38	.0007	.0005	.0003	.0002	.0002	.0001	.0001					0.38
0.46	.0013	.0009	.0007	.0005	.0003	.0002	.0002	.0001	.0001	.0001		0.46
0.54	.0024	.0017	.0012	.0008	.0006	.0004	.0003	.0002	.0001	.0001	.0001	0.54
0.62	.0039	.0028	.0020	.0014	.0010	.0007	.0005	.0004	.0003	.0002	.0001	0.62
0.70	.0060	.0043	.0031	.0022	.0016	.0012	.0008	.0006	.0004	.0003	.0002	0.70
0.78	.0088	.0064	.0046	.0033	.0024	.0018	.0013	.0009	.0007	.0005	.0004	0.78
0.86	.0123	.0090	.0066	.0048	.0035	.0026	.0019	.0014	.0010	.0007	.0005	0.86
0.94	.0166	.0122	.0090	.0066	.0049	.0036	.0026	.0019	.0014	.0011	.0008	0.94
1.02	.0218	.0162	.0120	.0089	.0066	.0049	.0036	.0027	.0020	.0015	.0011	1.02
1.10	.0279	.0209	.0156	.0117	.0088	.0066	.0049	.0037	.0027	.0021	.0015	1.10
1.18	.0351	.0265	.0200	.0151	.0114	.0086	.0065	.0049	.0037	.0028	.0021	1.18
1.26	.0433	.0329	.0250	.0190	.0145	.0110	.0084	.0064	.0048	.0037	.0028	1.26
1.34	.0526	.0403	.0309	.0237	.0181	.0139	.0107	.0082	.0063	.0048	.0037	1.34
1.42	.0630	.0487	.0376	.0291	.0225	.0173	.0134	.0104	.0080	.0062	.0048	1.42
1.50	.0746	.0581	.0452	.0352	.0274	.0214	.0166	.0130	.0101	.0079	.0061	1.50
1.58	.0873	.0686	.0538	.0423	.0332	.0260	.0204	.0160	.0126	.0099	.0078	1.58
1.66	.1013	.0801	.0634	.0502	.0397	.0314	.0249	.0197	.0156	.0123	.0098	1.66
1.74	.1164	.0928	.0741	.0591	.0471	.0376	.0300	.0239	.0191	.0152	.0121	1.74
1.82	.1327	.1067	.0858	.0690	.0555	.0446	.0359	.0289	.0232	.0187	.0150	1.82
1.90	.1503	.1218	.0987	.0800	.0649	.0526	.0426	.0346	.0280	.0227	.0184	1.90
1.98	.1690	.1381	.1129	.0922	.0753	.0616	.0503	.0411	.0336	.0274	.0224	1.98
2.06	.1890	.1557	.1282	.1056	.0870	.0716	.0590	.0486	.0400	.0330	.0272	2.06
2.14	.2102	.1745	.1449	.1203	.0999	.0829	.0688	.0572	.0475	.0394	.0327	2.14
2.22	.2325	.1946	.1629	.1363	.1141	.0955	.0799	.0669	.0560	.0468	.0392	2.22
2.30	.2560	.2160	.1822	.1537	.1297	.1094	.0923	.0779	.0657	.0554	.0468	2.30

Note: Blanks indicate P(>t) < 0.00005

Erlang-C Delay Loss Probabilities

(N = 5 Servers)

A	P(>0)	P(>t) for T_1/T_2 =										A
		0.1	0.2	0.3	0.4	0.5	0.6	0.7	0.8	0.9	1.0	
0.60	.0004	.0003	.0002	.0001	.0001							0.60
0.70	.0008	.0005	.0003	.0002	.0001	.0001	.0001					0.70
0.80	.0015	.0010	.0006	.0004	.0003	.0002	.0001	.0001	.0001			0.80
0.90	.0024	.0016	.0011	.0007	.0005	.0003	.0002	.0001	.0001	.0001		0.90
1.00	.0038	.0026	.0017	.0012	.0008	.0005	.0003	.0002	.0002	.0001	.0001	1.00
1.10	.0057	.0039	.0026	.0018	.0012	.0008	.0006	.0004	.0003	.0002	.0001	1.10
1.20	.0082	.0056	.0038	.0026	.0018	.0012	.0008	.0006	.0004	.0003	.0002	1.20
1.30	.0114	.0079	.0054	.0038	.0026	.0018	.0012	.0009	.0006	.0004	.0003	1.30
1.40	.0153	.0107	.0075	.0052	.0036	.0025	.0018	.0012	.0009	.0006	.0004	1.40
1.50	.0201	.0142	.0100	.0070	.0050	.0035	.0025	.0017	.0012	.0009	.0006	1.50
1.60	.0259	.0184	.0131	.0093	.0066	.0047	.0034	.0024	.0017	.0012	.0009	1.60
1.70	.0326	.0235	.0169	.0121	.0087	.0063	.0045	.0032	.0023	.0017	.0012	1.70
1.80	.0405	.0294	.0214	.0155	.0113	.0082	.0059	.0043	.0031	.0023	.0017	1.80
1.90	.0495	.0363	.0266	.0195	.0143	.0105	.0077	.0057	.0041	.0030	.0022	1.90
2.00	.0597	.0442	.0328	.0243	.0180	.0133	.0099	.0073	.0054	.0040	.0030	2.00
2.10	.0712	.0532	.0398	.0298	.0223	.0167	.0125	.0093	.0070	.0052	.0039	2.10
2.20	.0839	.0634	.0479	.0362	.0274	.0207	.0156	.0118	.0089	.0068	.0051	2.20
2.30	.0980	.0748	.0571	.0436	.0333	.0254	.0194	.0148	.0113	.0086	.0066	2.30
2.40	.1135	.0875	.0675	.0520	.0401	.0309	.0239	.0184	.0142	.0109	.0084	2.40
2.50	.1304	.1015	.0791	.0616	.0480	.0374	.0291	.0227	.0176	.0137	.0107	2.50
2.60	.1487	.1169	.0920	.0724	.0569	.0448	.0352	.0277	.0218	.0171	.0135	2.60
2.70	.1684	.1338	.1063	.0844	.0671	.0533	.0424	.0337	.0267	.0212	.0169	2.70
2.80	.1895	.1521	.1221	.0980	.0786	.0631	.0506	.0406	.0326	.0262	.0210	2.80
2.90	.2121	.1719	.1394	.1130	.0916	.0742	.0602	.0488	.0395	.0320	.0260	2.90
3.00	.2362	.1933	.1583	.1296	.1061	.0869	.0711	.0582	.0477	.0390	.0320	3.00

Note: Blanks indicate P(>t) < 0.00005

Erlang-C Delay Loss Probabilities

(N = 6 Servers)

A	P(>0)	P(>t) for T₁/T₂ =										A
		0.1	0.2	0.3	0.4	0.5	0.6	0.7	0.8	0.9	1.0	
1.60	.0064	.0041	.0027	.0017	.0011	.0007	.0005	.0003	.0002	.0001	.0001	1.60
1.70	.0085	.0056	.0036	.0024	.0015	.0010	.0006	.0004	.0003	.0002	.0001	1.70
1.80	.0111	.0073	.0048	.0032	.0021	.0014	.0009	.0006	.0004	.0003	.0002	1.80
1.90	.0143	.0095	.0063	.0042	.0028	.0018	.0012	.0008	.0005	.0004	.0002	1.90
2.00	.0180	.0121	.0081	.0054	.0036	.0024	.0016	.0011	.0007	.0005	.0003	2.00
2.10	.0224	.0152	.0103	.0070	.0047	.0032	.0022	.0015	.0010	.0007	.0005	2.10
2.20	.0275	.0188	.0129	.0088	.0060	.0041	.0028	.0019	.0013	.0009	.0006	2.20
2.30	.0333	.0230	.0159	.0110	.0076	.0052	.0036	.0025	.0017	.0012	.0008	2.30
2.40	.0400	.0279	.0194	.0136	.0095	.0066	.0046	.0032	.0022	.0016	.0011	2.40
2.50	.0474	.0334	.0236	.0166	.0117	.0082	.0058	.0041	.0029	.0020	.0014	2.50
2.60	.0558	.0397	.0283	.0201	.0143	.0102	.0073	.0052	.0037	.0026	.0019	2.60
2.70	.0652	.0468	.0337	.0242	.0174	.0125	.0090	.0065	.0047	.0033	.0024	2.70
2.80	.0755	.0548	.0398	.0289	.0210	.0152	.0111	.0080	.0058	.0042	.0031	2.80
2.90	.0868	.0637	.0467	.0342	.0251	.0184	.0135	.0099	.0073	.0053	.0039	2.90
3.00	.0991	.0734	.0544	.0403	.0299	.0221	.0164	.0121	.0090	.0067	.0049	3.00
3.10	.1126	.0842	.0630	.0472	.0353	.0264	.0198	.0148	.0111	.0083	.0062	3.10
3.20	.1271	.0961	.0726	.0549	.0415	.0313	.0237	.0179	.0135	.0102	.0077	3.20
3.30	.1427	.1090	.0832	.0635	.0485	.0370	.0282	.0216	.0165	.0126	.0096	3.30
3.40	.1595	.1230	.0948	.0731	.0564	.0435	.0335	.0258	.0199	.0154	.0118	3.40
3.50	.1775	.1382	.1076	.0838	.0653	.0508	.0396	.0308	.0240	.0187	.0146	3.50
3.60	.1966	.1546	.1216	.0957	.0753	.0592	.0466	.0366	.0288	.0227	.0178	3.60
3.70	.2168	.1723	.1369	.1088	.0864	.0687	.0546	.0433	.0344	.0274	.0217	3.70
3.80	.2383	.1912	.1535	.1232	.0988	.0793	.0637	.0511	.0410	.0329	.0264	3.80
3.90	.2609	.2115	.1714	.1390	.1126	.0913	.0740	.0600	.0486	.0394	.0320	3.90
4.00	.2848	.2331	.1909	.1563	.1280	.1048	.0858	.0702	.0575	.0471	.0385	4.00

Note: Blanks indicate P(>t) < 0.00005

Erlang-C Delay Loss Probabilities

(N = 7 Servers)

A	P(>0)	\(P(>t)\) for \(T_1/T_2 =\)										A
		0.1	0.2	0.3	0.4	0.5	0.6	0.7	0.8	0.9	1.0	
1.40	.0006	.0004	.0002	.0001	.0001							1.40
1.55	.0012	.0007	.0004	.0002	.0001	.0001						1.55
1.70	.0020	.0012	.0007	.0004	.0002	.0001	.0001					1.70
1.85	.0031	.0019	.0011	.0007	.0004	.0002	.0001	.0001	.0001			1.85
2.00	.0048	.0029	.0018	.0011	.0007	.0004	.0002	.0001	.0001	.0001		2.00
2.15	.0071	.0044	.0027	.0017	.0010	.0006	.0004	.0002	.0001	.0001	.0001	2.15
2.30	.0101	.0063	.0039	.0025	.0015	.0010	.0006	.0004	.0002	.0001	.0001	2.30
2.45	.0139	.0088	.0056	.0036	.0023	.0014	.0009	.0006	.0004	.0002	.0001	2.45
2.60	.0188	.0121	.0078	.0053	.0032	.0021	.0013	.0009	.0006	.0004	.0002	2.60
2.75	.0248	.0162	.0106	.0069	.0045	.0030	.0019	.0013	.0008	.0005	.0004	2.75
2.90	.0320	.0213	.0141	.0094	.0062	.0041	.0027	.0018	.0012	.0008	.0005	2.90
3.05	.0407	.0274	.0185	.0124	.0084	.0056	.0038	.0026	.0017	.0012	.0008	3.05
3.20	.0509	.0348	.0238	.0163	.0111	.0076	.0052	.0036	.0024	.0017	.0011	3.20
3.35	.0627	.0435	.0302	.0210	.0146	.0101	.0070	.0049	.0034	.0023	.0016	3.35
3.50	.0762	.0537	.0378	.0267	.0188	.0132	.0093	.0066	.0046	.0033	.0023	3.50
3.65	.0916	.0655	.0469	.0335	.0240	.0172	.0123	.0088	.0063	.0045	.0032	3.65
3.80	.1089	.0791	.0574	.0417	.0303	.0220	.0160	.0116	.0084	.0061	.0044	3.80
3.95	.1282	.0945	.0697	.0514	.0379	.0279	.0206	.0152	.0112	.0082	.0061	3.95
4.10	.1496	.1119	.0838	.0627	.0469	.0351	.0263	.0196	.0147	.0110	.0082	4.10
4.25	.1731	.1315	.0999	.0759	.0576	.0438	.0333	.0253	.0192	.0146	.0111	4.25
4.40	.1988	.1533	.1182	.0912	.0703	.0542	.0418	.0322	.0248	.0192	.0148	4.40
4.55	.2268	.1775	.1389	.1087	.0851	.0666	.0521	.0408	.0319	.0250	.0196	4.55
4.70	.2570	.2042	.1622	.1289	.1024	.0814	.0646	.0514	.0408	.0324	.0258	4.70
4.85	.2894	.2334	.1883	.1518	.1225	.0988	.0797	.0643	.0518	.0418	.0337	4.85
5.00	.3242	.2654	.2173	.1779	.1457	.1192	.0976	.0799	.0654	.0536	.0439	5.00

Note: Blanks indicate \(P(>t) < 0.00005\)

Erlang-C Delay Loss Probabilities

(N = 8 Servers)

| A | P(>0) | \multicolumn{11}{c}{P(>t) for T_1/T_2 =} | A |
		0.1	0.2	0.3	0.4	0.5	0.6	0.7	0.8	0.9	1.0	
1.4	.0001	.0001	.0001										1.4
1.6	.0003	.0001	.0002										1.6
1.8	.0006	.0003	.0002	.0001									1.8
2.0	.0011	.0006	.0003	.0002	.0001								2.0
2.2	.0021	.0012	.0007	.0004	.0002	.0001							2.2
2.4	.0035	.0020	.0012	.0007	.0004	.0002	.0001	.0001					2.4
2.6	.0057	.0033	.0019	.0011	.0007	.0004	.0002	.0001	.0001				2.6
2.8	.0088	.0052	.0031	.0018	.0011	.0007	.0004	.0002	.0001	.0001			2.8
3.0	.0129	.0079	.0048	.0029	.0018	.0011	.0006	.0004	.0002	.0001	.0001		3.0
3.2	.0185	.0114	.0071	.0044	.0027	.0017	.0010	.0006	.0004	.0002	.0002		3.2
3.4	.0256	.0162	.0102	.0065	.0041	.0026	.0016	.0010	.0006	.0004	.0003		3.4
3.6	.0346	.0223	.0144	.0092	.0060	.0038	.0025	.0016	.0010	.0007	.0004		3.6
3.8	.0457	.0300	.0197	.0130	.0085	.0056	.0037	.0024	.0016	.0010	.0007		3.8
4.0	.0590	.0396	.0265	.0178	.0119	.0080	.0054	.0036	.0024	.0016	.0011		4.0
4.2	.0749	.0512	.0350	.0240	.0164	.0112	.0077	.0052	.0036	.0025	.0017		4.2
4.4	.0935	.0652	.0455	.0318	.0222	.0155	.0108	.0075	.0052	.0037	.0026		4.4
4.6	.1150	.0819	.0583	.0415	.0295	.0210	.0150	.0106	.0076	.0054	.0038		4.6
4.8	.1395	.1013	.0736	.0534	.0388	.0282	.0205	.0149	.0108	.0078	.0057		4.8
5.0	.1673	.1239	.0918	.0680	.0504	.0373	.0276	.0205	.0152	.0112	.0083		5.0
5.2	.1983	.1499	.1133	.0856	.0647	.0489	.0370	.0279	.0211	.0160	.0121		5.2
5.4	.2327	.1794	.1384	.1067	.0823	.0634	.0489	.0377	.0291	.0224	.0173		5.4
5.6	.2706	.2129	.1674	.1317	.1036	.0815	.0641	.0504	.0397	.0312	.0245		5.6
5.8	.3120	.2504	.2009	.1613	.1294	.1039	.0833	.0669	.0537	.0431	.0346		5.8
6.0	.3570	.2923	.2393	.1959	.1604	.1313	.1075	.0880	.0721	.0590	.0483		6.0
6.2	.4055	.3387	.2829	.2363	.1974	.1649	.1377	.1150	.0961	.0803	.0670		6.2

Note: Blanks indicate P(>t) < 0.00005

Erlang-C Delay Loss Probabilities

(N = 9 Servers)

P(>t) for T$_1$/T$_2$ =

A	P(>0)	0.1	0.2	0.3	0.4	0.5	0.6	0.7	0.8	0.9	1.0	A
2.2	.0005	.0002	.0001									2.2
2.4	.0009	.0005	.0002	.0001								2.4
2.6	.0016	.0008	.0004	.0002	.0001							2.6
2.8	.0026	.0014	.0007	.0004	.0002	.0001						2.8
3.0	.0040	.0022	.0012	.0007	.0004	.0002	.0001					3.0
3.2	.0061	.0034	.0019	.0011	.0006	.0003	.0002	.0001				3.2
3.4	.0090	.0051	.0029	.0017	.0010	.0005	.0003	.0002	.0001			3.4
3.6	.0127	.0074	.0043	.0025	.0015	.0009	.0005	.0003	.0002	.0001		3.6
3.8	.0176	.0105	.0062	.0037	.0022	.0013	.0008	.0005	.0003	.0002	.0001	3.8
4.0	.0238	.0144	.0087	.0053	.0032	.0020	.0012	.0007	.0004	.0003	.0002	4.0
4.2	.0314	.0192	.0120	.0074	.0046	.0028	.0018	.0011	.0007	.0004	.0003	4.2
4.4	.0407	.0257	.0162	.0102	.0065	.0041	.0026	.0016	.0010	.0006	.0004	4.4
4.6	.0519	.0334	.0215	.0139	.0089	.0057	.0037	.0024	.0015	.0010	.0006	4.6
4.8	.0651	.0428	.0281	.0185	.0121	.0080	.0052	.0034	.0023	.0015	.0010	4.8
5.0	.0805	.0540	.0362	.0242	.0163	.0109	.0073	.0049	.0033	.0022	.0015	5.0
5.2	.0983	.0672	.0460	.0314	.0215	.0147	.0101	.0069	.0047	.0032	.0022	5.2
5.4	.1186	.0828	.0577	.0403	.0281	.0196	.0137	.0095	.0067	.0046	.0032	5.4
5.6	.1416	.1008	.0717	.0511	.0363	.0259	.0184	.0131	.0093	.0066	.0047	5.6
5.8	.1673	.1215	.0882	.0641	.0465	.0338	.0245	.0178	.0129	.0094	.0068	5.8
6.0	.1960	.1452	.1076	.0797	.0590	.0437	.0324	.0240	.0178	.0132	.0098	6.0
6.2	.2276	.1720	.1300	.0982	.0743	.0561	.0424	.0321	.0242	.0183	.0138	6.2
6.4	.2622	.2022	.1559	.1202	.0927	.0715	.0551	.0425	.0328	.0253	.0195	6.4
6.6	.3000	.2367	.1856	.1460	.1149	.0904	.0711	.0559	.0440	.0346	.0272	6.6
6.8	.3409	.2736	.2195	.1762	.1414	.1135	.0911	.0731	.0586	.0471	.0378	6.8
7.0	.3849	.3152	.2580	.2113	.1730	.1416	.1159	.0949	.0777	.0636	.0521	7.0

Note: Blanks indicate P(>t) < 0.00005

Erlang-C Delay Loss Probabilities

(N = 10 Servers)

A	P(>0)	\$P(>t)\$ for \$T_1/T_2\$ =										A
		0.1	0.2	0.3	0.4	0.5	0.6	0.7	0.8	0.9	1.0	
3.2	.0019	.0009	.0005	.0002	.0001	.0001						3.2
3.4	.0029	.0015	.0008	.0004	.0002	.0001	.0001					3.4
3.6	.0043	.0023	.0012	.0006	.0003	.0002	.0001					3.6
3.8	.0062	.0034	.0018	.0010	.0005	.0003	.0002	.0001				3.8
4.0	.0088	.0048	.0027	.0015	.0008	.0004	.0002	.0001	.0001			4.0
4.2	.0122	.0068	.0038	.0021	.0012	.0007	.0004	.0002	.0001	.0001		4.2
4.4	.0164	.0094	.0054	.0031	.0017	.0010	.0006	.0003	.0002	.0001	.0001	4.4
4.6	.0217	.0127	.0074	.0043	.0025	.0015	.0009	.0005	.0003	.0002	.0001	4.6
4.8	.0282	.0168	.0100	.0059	.0035	.0021	.0012	.0007	.0004	.0003	.0002	4.8
5.0	.0361	.0219	.0133	.0081	.0049	.0030	.0018	.0011	.0007	.0004	.0002	5.0
5.2	.0455	.0282	.0174	.0108	.0067	.0041	.0026	.0016	.0010	.0006	.0004	5.2
5.4	.0566	.0357	.0225	.0142	.0090	.0057	.0036	.0023	.0014	.0009	.0006	5.4
5.6	.0695	.0447	.0288	.0186	.0120	.0077	.0050	.0032	.0021	.0013	.0009	5.6
5.8	.0843	.0554	.0364	.0239	.0157	.0103	.0068	.0045	.0029	.0019	.0013	5.8
6.0	.1013	.0679	.0455	.0305	.0205	.0137	.0092	.0062	.0041	.0028	.0019	6.0
6.2	.1205	.0824	.0564	.0385	.0264	.0180	.0123	.0084	.0058	.0039	.0027	6.2
6.4	.1420	.0991	.0691	.0482	.0337	.0235	.0164	.0114	.0080	.0056	.0039	6.4
6.6	.1660	.1182	.0841	.0599	.0426	.0303	.0216	.0154	.0109	.0078	.0055	6.6
6.8	.1926	.1398	.1015	.0737	.0535	.0389	.0282	.0205	.0149	.0108	.0078	6.8
7.0	.2217	.1643	.1217	.0901	.0668	.0495	.0367	.0272	.0201	.0149	.0110	7.0
7.2	.2536	.1917	.1449	.1095	.0827	.0625	.0473	.0357	.0270	.0204	.0154	7.2
7.4	.2882	.2222	.1714	.1321	.1019	.0786	.0606	.0467	.0360	.0278	.0214	7.4
7.6	.3257	.2562	.2015	.1585	.1247	.0981	.0772	.0607	.0477	.0376	.0295	7.6
7.8	.3660	.2937	.2357	.1892	.1518	.1218	.0978	.0785	.0630	.0505	.0406	7.8
8.0	.4092	.3350	.2743	.2246	.1839	.1505	.1232	.1009	.0826	.0676	.0554	8.0

Note: Blanks indicate \$P(>t) < 0.00005\$

Erlang-C Delay Loss Probabilities

(N = 11 Servers)

A	P(>0)	P(>t) for T_1/T_2 =										A
		0.1	0.2	0.3	0.4	0.5	0.6	0.7	0.8	0.9	1.0	
4.2	.0044	.0022	.0011	.0006	.0003	.0001	.0001					4.2
4.4	.0061	.0032	.0016	.0008	.0004	.0002	.0001	.0001				4.4
4.6	.0084	.0045	.0023	.0012	.0007	.0003	.0002	.0001	.0001			4.6
4.8	.0114	.0061	.0033	.0018	.0010	.0005	.0003	.0001	.0001			4.8
5.0	.0151	.0083	.0045	.0025	.0014	.0008	.0004	.0002	.0001	.0001		5.0
5.2	.0197	.0110	.0062	.0035	.0019	.0011	.0006	.0003	.0002	.0001	.0001	5.2
5.4	.0252	.0144	.0082	.0047	.0027	.0015	.0009	.0005	.0003	.0002	.0001	5.4
5.6	.0319	.0186	.0108	.0063	.0037	.0021	.0013	.0007	.0004	.0002	.0001	5.6
5.8	.0399	.0237	.0141	.0084	.0050	.0030	.0018	.0010	.0006	.0004	.0002	5.8
6.0	.0492	.0299	.0181	.0110	.0067	.0040	.0025	.0015	.0009	.0005	.0003	6.0
6.2	.0601	.0372	.0230	.0142	.0088	.0055	.0034	.0021	.0013	.0008	.0005	6.2
6.4	.0762	.0458	.0289	.0183	.0115	.0073	.0046	.0029	.0018	.0012	.0007	6.4
6.6	.0868	.0559	.0360	.0232	.0149	.0096	.0062	.0040	.0026	.0017	.0011	6.6
6.8	.1030	.0677	.0445	.0292	.0192	.0126	.0083	.0054	.0036	.0024	.0015	6.8
7.0	.1211	.0812	.0544	.0365	.0245	.0164	.0110	.0074	.0049	.0033	.0022	7.0
7.2	.1413	.0967	.0661	.0452	.0309	.0211	.0145	.0099	.0068	.0046	.0032	7.2
7.4	.1638	.1142	.0797	.0556	.0388	.0271	.0189	.0132	.0092	.0064	.0045	7.4
7.6	.1884	.1341	.0955	.0680	.0484	.0344	.0245	.0174	.0124	.0088	.0063	7.6
7.8	.2155	.1565	.1136	.0825	.0599	.0435	.0316	.0229	.0167	.0121	.0088	7.8
8.0	.2450	.1815	.1344	.0996	.0738	.0547	.0405	.0300	.0222	.0165	.0122	8.0
8.2	.2769	.2093	.1582	.1195	.0904	.0683	.0516	.0390	.0285	.0223	.0168	8.2
8.4	.3114	.2401	.1851	.1427	.1101	.0849	.0654	.0505	.0389	.0300	.0231	8.4
8.6	.3485	.2741	.2156	.1696	.1334	.1050	.0826	.0649	.0511	.0402	.0316	8.6
8.8	.3881	.3115	.2500	.2006	.1610	.1292	.1037	.0832	.0668	.0536	.0430	8.8
9.0	.4305	.3524	.2886	.2362	.1934	.1584	.1297	.1062	.0869	.0712	.0583	9.0

Note: Blanks indicate P(>t) < 0.00005

Erlang-C Delay Loss Probabilities

(N = 12 Servers)

A	P(>0)	P(>t) for T_1/T_2 =										A
		0.1	0.2	0.3	0.4	0.5	0.6	0.7	0.8	0.9	1.0	
5.0	.0059	.0029	.0015	.0007	.0004	.0002	.0001	.0001				5.0
5.2	.0079	.0040	.0020	.0010	.0005	.0003	.0001	.0001				5.2
5.4	.0105	.0054	.0028	.0015	.0008	.0004	.0002	.0002	.0001			5.4
5.6	.0138	.0073	.0038	.0020	.0011	.0006	.0003	.0002	.0001			5.6
5.8	.0177	.0095	.0051	.0028	.0015	.0008	.0004	.0002	.0001	.0001		5.8
6.0	.0225	.0123	.0068	.0037	.0020	.0011	.0006	.0003	.0002	.0001	.0001	6.0
6.2	.0282	.0158	.0088	.0049	.0028	.0016	.0009	.0005	.0003	.0002	.0001	6.2
6.4	.0349	.0200	.0114	.0065	.0037	.0021	.0012	.0007	.0004	.0002	.0001	6.4
6.6	.0429	.0250	.0146	.0085	.0049	.0029	.0017	.0010	.0006	.0003	.0002	6.6
6.8	.0521	.0309	.0184	.0109	.0065	.0039	.0023	.0014	.0008	.0005	.0003	6.8
7.0	.0626	.0380	.0230	.0140	.0085	.0051	.0031	.0019	.0011	.0007	.0004	7.0
7.2	.0747	.0462	.0286	.0177	.0109	.0068	.0042	.0026	.0016	.0010	.0006	7.2
7.4	.0883	.0558	.0352	.0222	.0140	.0089	.0056	.0035	.0022	.0014	.0009	7.4
7.6	.1037	.0668	.0430	.0277	.0178	.0115	.0074	.0048	.0031	.0020	.0013	7.6
7.8	.1208	.0794	.0522	.0343	.0225	.0148	.0097	.0064	.0042	.0028	.0018	7.8
8.0	.1398	.0937	.0628	.0421	.0282	.0189	.0127	.0085	.0057	.0038	.0026	8.0
8.2	.1608	.1100	.0752	.0514	.0352	.0241	.0165	.0113	.0077	.0053	.0036	8.2
8.4	.1839	.1283	.0895	.0624	.0436	.0304	.0212	.0148	.0103	.0072	.0050	8.4
8.6	.2091	.1488	.1059	.0754	.0537	.0382	.0272	.0193	.0138	.0098	.0070	8.6
8.8	.2364	.1717	.1247	.0905	.0657	.0477	.0347	.0252	.0183	.0133	.0096	8.8
9.0	.2660	.1971	.1460	.1082	.0801	.0594	.0440	.0326	.0241	.0179	.0132	9.0
9.2	.2979	.2252	.1702	.1286	.0972	.0735	.0555	.0420	.0317	.0240	.0181	9.2
9.4	.3322	.2561	.1975	.1523	.1174	.0905	.0698	.0538	.0415	.0320	.0247	9.4
9.6	.3688	.2901	.2282	.1795	.1412	.1111	.0874	.0687	.0541	.0425	.0335	9.6
9.8	.4079	.3273	.2627	.2108	.1692	.1358	.1090	.0874	.0702	.0563	.0452	9.8

Note: Blanks indicate P(>t) < 0.00005

Erlang-C Delay Loss Probabilities

(N = 13 Servers)

A	P(>0)	0.1	0.2	0.3	0.4	0.5	0.6	0.7	0.8	0.9	1.0	A
						P(>t) for T_1/T_2 =						
6.0	.0096	.0048	.0024	.0012	.0006	.0003	.0001	.0001	.0001			6.0
6.2	.0124	.0063	.0032	.0016	.0008	.0004	.0002	.0001	.0001			6.2
6.4	.0159	.0082	.0042	.0022	.0011	.0006	.0003	.0002	.0001	.0001		6.4
6.6	.0200	.0105	.0056	.0029	.0015	.0008	.0004	.0002	.0001	.0001	.0001	6.6
6.8	.0249	.0134	.0072	.0039	.0021	.0011	.0006	.0003	.0002	.0001	.0001	6.8
7.0	.0306	.0168	.0092	.0051	.0028	.0015	.0008	.0005	.0003	.0001	.0001	7.0
7.2	.0374	.0209	.0117	.0066	.0037	.0021	.0012	.0006	.0004	.0002	.0001	7.2
7.4	.0452	.0258	.0147	.0084	.0048	.0027	.0016	.0009	.0005	.0003	.0002	7.4
7.6	.0542	.0316	.0184	.0107	.0062	.0036	.0021	.0012	.0007	.0004	.0002	7.6
7.8	.0644	.0383	.0228	.0135	.0080	.0048	.0028	.0017	.0010	.0006	.0004	7.8
8.0	.0760	.0461	.0280	.0170	.0103	.0062	.0038	.0023	.0014	.0008	.0005	8.0
8.2	.0890	.0551	.0341	.0211	.0131	.0081	.0050	.0031	.0019	.0012	.0007	8.2
8.4	.1036	.0654	.0413	.0261	.0165	.0104	.0066	.0041	.0026	.0017	.0010	8.4
8.6	.1199	.0772	.0497	.0320	.0206	.0133	.0086	.0055	.0035	.0023	.0015	8.6
8.8	.1378	.0905	.0595	.0391	.0257	.0169	.0111	.0073	.0048	.0031	.0021	8.8
9.0	.1575	.1056	.0708	.0474	.0318	.0213	.0143	.0096	.0064	.0043	.0029	9.0
9.2	.1791	.1225	.0837	.0573	.0392	.0268	.0183	.0125	.0086	.0059	.0040	9.2
9.4	.2026	.1413	.0986	.0688	.0480	.0335	.0234	.0163	.0114	.0079	.0055	9.4
9.6	.2281	.1623	.1155	.0822	.0585	.0417	.0297	.0211	.0150	.0107	.0076	9.6
9.8	.2556	.1856	.1348	.0979	.0711	.0516	.0375	.0272	.0198	.0143	.0104	9.8
10.0	.2853	.2113	.1566	.1160	.0859	.0637	.0472	.0349	.0259	.0192	.0142	10.0
10.2	.3171	.2396	.1811	.1369	.1034	.0782	.0591	.0447	.0338	.0255	.0193	10.2
10.4	.3510	.2707	.2087	.1609	.1241	.0957	.0738	.0569	.0439	.0338	.0261	10.4
10.6	.3872	.3046	.2396	.1885	.1483	.1166	.0917	.0722	.0568	.0447	.0351	10.6
10.8	.4257	.3416	.2741	.2200	.1766	.1417	.1137	.0913	.0732	.0588	.0472	10.8

Note: Blanks indicate P(>t) < 0.00005

Erlang-C Delay Loss Probabilities

(N = 14 Servers)

A	P(>0)	P(>t) for T₁/T₂ =										A

$$P(>t) \text{ for } T_1/T_2 =$$

A	P(>0)	0.1	0.2	0.3	0.4	0.5	0.6	0.7	0.8	0.9	1.0	A
7.0	.0142	.0070	.0035	.0017	.0009	.0004	.0002	.0001	.0001			7.0
7.2	.0177	.0090	.0045	.0023	.0012	.0006	.0003	.0002	.0001			7.2
7.4	.0219	.0113	.0059	.0030	.0016	.0008	.0004	.0002	.0001	.0001		7.4
7.6	.0268	.0142	.0075	.0039	.0021	.0011	.0006	.0003	.0002	.0001		7.6
7.8	.0326	.0175	.0094	.0051	.0027	.0015	.0008	.0004	.0002	.0001	.0001	7.8
8.0	.0393	.0216	.0118	.0065	.0036	.0020	.0011	.0006	.0003	.0002	.0001	8.0
8.2	.0469	.0263	.0147	.0082	.0046	.0026	.0014	.0008	.0005	.0003	.0001	8.2
8.4	.0557	.0318	.0182	.0104	.0059	.0034	.0019	.0011	.0006	.0004	.0002	8.4
8.6	.0656	.0382	.0223	.0130	.0076	.0044	.0026	.0015	.0009	.0005	.0003	8.6
8.8	.0767	.0456	.0271	.0161	.0096	.0057	.0034	.0020	.0012	.0007	.0004	8.8
9.0	.0892	.0541	.0328	.0199	.0121	.0073	.0044	.0027	.0016	.0010	.0006	9.0
9.2	.1031	.0638	.0395	.0244	.0151	.0093	.0058	.0036	.0022	.0014	.0008	9.2
9.4	.1184	.0748	.0472	.0298	.0188	.0119	.0075	.0047	.0030	.0019	.0012	9.4
9.6	.1353	.0872	.0561	.0362	.0233	.0150	.0097	.0062	.0040	.0026	.0017	9.6
9.8	.1539	.1011	.0664	.0436	.0287	.0188	.0124	.0081	.0053	.0035	.0023	9.8
10.0	.1741	.1167	.0782	.0524	.0352	.0236	.0158	.0106	.0071	.0048	.0032	10.0
10.2	.1961	.1341	.0917	.0627	.0429	.0293	.0201	.0137	.0094	.0064	.0044	10.2
10.4	.2200	.1535	.1071	.0747	.0521	.0364	.0254	.0177	.0123	.0086	.0060	10.4
10.6	.2457	.1749	.1245	.0886	.0631	.0449	.0319	.0227	.0162	.0115	.0082	10.6
10.8	.2733	.1985	.1441	.1047	.0760	.0552	.0401	.0291	.0211	.0153	.0111	10.8
11.0	.3029	.2244	.1662	.1232	.0912	.0676	.0501	.0371	.0275	.0204	.0151	11.0
11.2	.3345	.2528	.1911	.1444	.1092	.0825	.0623	.0471	.0356	.0269	.0203	11.2
11.4	.3682	.2839	.2189	.1688	.1301	.1003	.0774	.0597	.0460	.0355	.0273	11.4
11.6	.4039	.3177	.2499	.1966	.1547	.1217	.0957	.0753	.0592	.0466	.0366	11.6
11.8	.4417	.3545	.2845	.2283	.1832	.1470	.1180	.0947	.0760	.0610	.0489	11.8

Note: Blanks indicate P(>t) < 0.00005

Erlang-C Delay Loss Probabilities

(N = 15 Servers)

A	P(>0)	\multicolumn{11}{c}{P(>t) for T_1/T_2 =}											A
		0.1	0.2	0.3	0.4	0.5	0.6	0.7	0.8	0.9	1.0	
8.0	.0193	.0096	.0048	.0024	.0012	.0006	.0003	.0001	.0001			8.0
8.2	.0235	.0119	.0060	.0031	.0016	.0008	.0004	.0002	.0001	.0001		8.2
8.4	.0285	.0147	.0076	.0039	.0020	.0011	.0005	.0003	.0001	.0001	.0001	8.4
8.6	.0342	.0180	.0095	.0050	.0026	.0014	.0007	.0004	.0002	.0001	.0001	8.6
8.8	.0408	.0219	.0118	.0063	.0034	.0018	.0010	.0005	.0003	.0002	.0001	8.8
9.0	.0482	.0265	.0145	.0080	.0044	.0024	.0013	.0007	.0004	.0002	.0001	9.0
9.2	.0567	.0317	.0178	.0100	.0056	.0031	.0017	.0010	.0005	.0003	.0002	9.2
9.4	.0662	.0378	.0216	.0123	.0071	.0040	.0023	.0013	.0008	.0004	.0002	9.4
9.6	.0769	.0448	.0261	.0152	.0089	.0052	.0030	.0018	.0010	.0006	.0003	9.6
9.8	.0888	.0528	.0314	.0187	.0111	.0066	.0039	.0023	.0014	.0008	.0005	9.8
10.0	.1020	.0619	.0375	.0228	.0138	.0084	.0051	.0031	.0019	.0011	.0007	10.0
10.2	.1166	.0722	.0446	.0276	.0171	.0106	.0065	.0041	.0025	.0016	.0010	10.2
10.4	.1326	.0837	.0528	.0334	.0211	.0133	.0084	.0053	.0033	.0021	.0013	10.4
10.6	.1501	.0967	.0623	.0401	.0258	.0166	.0107	.0069	.0044	.0029	.0018	10.6
10.8	.1691	.1111	.0730	.0480	.0315	.0207	.0136	.0089	.0059	.0039	.0025	10.8
11.0	.1898	.1272	.0853	.0572	.0383	.0257	.0172	.0115	.0077	.0052	.0035	11.0
11.2	.2121	.1451	.0992	.0678	.0464	.0317	.0217	.0148	.0101	.0069	.0047	11.2
11.4	.2362	.1648	.1150	.0802	.0560	.0390	.0272	.0190	.0133	.0093	.0065	11.4
11.6	.2620	.1865	.1328	.0945	.0673	.0479	.0341	.0243	.0173	.0123	.0087	11.6
11.8	.2897	.2104	.1527	.1109	.0805	.0585	.0425	.0308	.0224	.0163	.0118	11.8
12.0	.3192	.2365	.1752	.1298	.0961	.0712	.0528	.0391	.0290	.0215	.0159	12.0
12.2	.3506	.2650	.2003	.1514	.1144	.0865	.0653	.0494	.0373	.0282	.0213	12.2
12.4	.3839	.2960	.2282	.1760	.1357	.1046	.0807	.0622	.0480	.0370	.0285	12.4
12.6	.4192	.3297	.2594	.2040	.1605	.1263	.0993	.0781	.0615	.0483	.0380	12.6
12.8	.4564	.3663	.2940	.2359	.1893	.1519	.1219	.0979	.0785	.0630	.0506	12.8

Note: Blanks indicate P(>t) < 0.00005

Erlang-C Delay Loss Probabilities

(N = 16 Servers)

A	P(>0)	\multicolumn{10}{c	}{P(>t) for T1/T2 =}	A
		0.1	0.2	0.3	0.4	0.5	0.6	0.7	0.8	0.9	1.0	
9.0	.0249	.0124	.0061	.0031	.0015	.0008	.0004	.0002	.0001			9.0
9.2	.0298	.0151	.0077	.0039	.0020	.0010	.0005	.0003	.0001			9.2
9.4	.0354	.0183	.0095	.0049	.0025	.0013	.0007	.0003	.0002			9.4
9.6	.0419	.0221	.0116	.0061	.0032	.0017	.0009	.0005	.0003	.0001	.0001	9.6
9.8	.0491	.0264	.0142	.0077	.0041	.0022	.0012	.0006	.0003	.0002	.0001	9.8
10.0	.0573	.0315	.0173	.0095	.0052	.0029	.0016	.0009	.0005	.0003	.0001	10.0
10.2	.0665	.0372	.0209	.0117	.0065	.0037	.0020	.0011	.0006	.0004	.0002	10.2
10.4	.0768	.0439	.0251	.0143	.0082	.0047	.0027	.0015	.0009	.0005.	.0003	10.4
10.6	.0882	.0514	.0299	.0174	.0102	.0059	.0035	.0020	.0012	.0007	.0004	10.6
10.8	.1007	.0599	.0356	.0212	.0126	.0075	.0044	.0026	.0016	.0009	.0006	10.8
11.0	.1145	.0695	.0421	.0256	.0155	.0094	.0057	.0035	.0021	.0013	.0008	11.0
11.2	.1297	.0802	.0497	.0307	.0190	.0118	.0073	.0045	.0028	.0017	.0011	11.2
11.4	.1462	.0923	.0583	.0368	.0232	.0147	.0093	.0058	.0037	.0023	.0015	11.4
11.6	.1641	.1057	.0681	.0438	.0282	.0182	.0117	.0075	.0049	.0031	.0020	11.6
11.8	.1836	.1206	.0793	.0521	.0342	.0225	.0148	.0097	.0064	.0042	.0028	11.8
12.0	.2046	.1371	.0919	.0616	.0413	.0277	.0186	.0124	.0083	.0056	.0037	12.0
12.2	.2272	.1553	.1062	.0726	.0497	.0340	.0232	.0159	.0109	.0074	.0051	12.2
12.4	.2514	.1754	.1224	.0854	.0596	.0416	.0290	.0202	.0141	.0098	.0069	12.4
12.6	.2773	.1973	.1405	.1000	.0712	.0507	.0361	.0257	.0183	.0131	.0093	12.6
12.8	.3049	.2214	.1608	.1167	.0848	.0616	.0447	.0325	.0236	.0171	.0124	12.8
13.0	.3343	.2476	.1834	.1359	.1007	.0746	.0553	.0409	.0303	.0225	.0166	13.0
13.2	.3654	.2762	.2087	.1578	.1192	.0901	.0681	.0515	.0389	.0294	.0222	13.2
13.4	.3984	.3072	.2369	.1826	.1408	.1086	.0837	.0646	.0498	.0384	.0296	13.4
13.6	.4332	.3408	.2681	.2109	.1659	.1305	.1026	.0807	.0635	.0500	.0393	13.6
13.8	.4699	.3771	.3026	.2429	.1949	.1564	.1255	.1007	.0808	.0649	.0521	13.8

Note: Blanks indicate P(>t) < 0.00005

Erlang-C Delay Loss Probabilities

(N = 17 Servers)

A	P(>0)	P(>t) for T_1/T_2 =										A
		0.1	0.2	0.3	0.4	0.5	0.6	0.7	0.8	0.9	1.0	
10.0	.0309	.0153	.0076	.0038	.0019	.0009	.0005	.0002	.0001	.0001		10.0
10.2	.0364	.0184	.0093	.0047	.0024	.0012	.0006	.0003	.0002	.0001		10.2
10.4	.0427	.0221	.0114	.0059	.0030	.0016	.0008	.0004	.0002	.0001	.0001	10.4
10.6	.0497	.0262	.0138	.0073	.0038	.0020	.0011	.0006	.0003	.0002	.0001	10.6
10.8	.0577	.0310	.0167	.0090	.0048	.0026	.0014	.0008	.0004	.0002	.0001	10.8
11.0	.0665	.0365	.0200	.0110	.0060	.0033	.0018	.0010	.0005	.0003	.0002	11.0
11.2	.0763	.0427	.0239	.0134	.0075	.0042	.0024	.0013	.0007	.0004	.0002	11.2
11.4	.0872	.0498	.0284	.0162	.0093	.0053	.0030	.0017	.0010	.0006	.0003	11.4
11.6	.0992	.0578	.0337	.0196	.0114	.0067	.0039	.0023	.0013	.0008	.0004	11.6
11.8	.1123	.0668	.0397	.0236	.0140	.0083	.0050	.0029	.0018	.0010	.0006	11.8
12.0	.1266	.0768	.0466	.0283	.0171	.0104	.0063	.0038	.0023	.0014	.0009	12.0
12.2	.1423	.0880	.0545	.0337	.0209	.0129	.0080	.0049	.0031	.0019	.0012	12.2
12.4	.1592	.1005	.0634	.0401	.0253	.0160	.0101	.0064	.0040	.0025	.0016	12.4
12.6	.1775	.1143	.0736	.0474	.0305	.0197	.0127	.0082	.0053	.0034	.0022	12.6
12.8	.1973	.1296	.0852	.0560	.0368	.0242	.0159	.0104	.0069	.0045	.0030	12.8
13.0	.2185	.1465	.0982	.0658	.0441	.0296	.0198	.0133	.0089	.0060	.0040	13.0
13.2	.2413	.1650	.1128	.0772	.0528	.0361	.0247	.0169	.0115	.0079	.0054	13.2
13.4	.2656	.1853	.1293	.0902	.0629	.0439	.0306	.0214	.0149	.0104	.0073	13.4
13.6	.2915	.2075	.1477	.1051	.0748	.0533	.0379	.0270	.0192	.0137	.0097	13.6
13.8	.3191	.2317	.1682	.1222	.0887	.0644	.0468	.0340	.0247	.0179	.0130	13.8
14.0	.3483	.2580	.1911	.1416	.1049	.0777	.0576	.0426	.0316	.0234	.0173	14.0
14.2	.3792	.2866	.2166	.1637	.1237	.0935	.0707	.0534	.0404	.0305	.0231	14.2
14.4	.4118	.3175	.2448	.1888	.1456	.1122	.0865	.0667	.0514	.0397	.0306	14.4
14.6	.4462	.3510	.2761	.2172	.1708	.1344	.1057	.0832	.0654	.0515	.0405	14.6
14.8	.4823	.3871	.3106	.2493	.2001	.1606	.1288	.1034	.0830	.0666	.0534	14.8

Note: Blanks indicate P(>t) < 0.00005

Erlang-C Delay Loss Probabilities

(N = 18 Servers)

A	P(>0)	\multicolumn{11}{c}{P(>t) for T₁/T₂ =}	A

A	P(>0)	0.1	0.2	0.3	0.4	0.5	0.6	0.7	0.8	0.9	1.0	A
11.0	.0371	.0184	.0092	.0045	.0023	.0011	.0006	.0003	.0001	.0001		11.0
11.2	.0432	.0219	.0111	.0056	.0028	.0014	.0007	.0004	.0002	.0001		11.2
11.4	.0500	.0259	.0134	.0069	.0036	.0018	.0010	.0005	.0003	.0001	.0001	11.4
11.6	.0577	.0304	.0160	.0085	.0045	.0024	.0012	.0007	.0003	.0002	.0001	11.6
11.8	.0662	.0356	.0192	.0103	.0055	.0030	.0016	.0009	.0005	.0002	.0001	11.8
12.0	.0756	.0415	.0228	.0125	.0069	.0033	.0021	.0011	.0006	.0003	.0002	12.0
12.2	.0860	.0482	.0270	.0151	.0085	.0047	.0026	.0015	.0008	.0005	.0003	12.2
12.4	.0974	.0556	.0318	.0182	.0104	.0059	.0034	.0019	.0011	.0006	.0004	12.4
12.6	.1099	.0640	.0373	.0217	.0127	.0074	.0043	.0025	.0015	.0009	.0005	12.6
12.8	.1235	.0734	.0437	.0260	.0154	.0092	.0055	.0032	.0019	.0011	.0007	12.8
13.0	.1383	.0839	.0509	.0309	.0187	.0114	.0069	.0042	.0025	.0015	.0009	13.0
13.2	.1543	.0955	.0591	.0366	.0226	.0140	.0087	.0054	.0033	.0021	.0013	13.2
13.4	.1716	.1084	.0684	.0432	.0273	.0172	.0109	.0069	.0043	.0027	.0017	13.4
13.6	.1903	.1226	.0789	.0508	.0327	.0211	.0136	.0087	.0056	.0036	.0023	13.6
13.8	.2103	.1382	.0908	.0597	.0392	.0258	.0169	.0111	.0073	.0048	.0032	13.8
14.0	.2317	.1553	.1041	.0698	.0468	.0314	.0210	.0141	.0094	.0063	.0042	14.0
14.2	.2546	.1741	.1191	.0814	.0557	.0381	.0260	.0178	.0122	.0083	.0057	14.2
14.4	.2790	.1946	.1358	.0947	.0661	.0461	.0322	.0224	.0157	.0109	.0076	14.4
14.6	.3049	.2170	.1544	.1099	.0782	.0557	.0396	.0282	.0201	.0143	.0102	14.6
14.8	.3323	.2413	.1752	.1272	.0924	.0671	.0487	.0354	.0257	.0187	.0135	14.8
15.0	.3613	.2677	.1983	.1469	.1088	.0806	.0597	.0442	.0328	.0243	.0180	15.0
15.2	.3920	.2963	.2239	.1692	.1279	.0967	.0731	.0552	.0417	.0315	.0238	15.2
15.4	.4243	.3271	.2522	.1945	.1500	.1156	.0892	.0687	.0530	.0409	.0315	15.4
15.6	.4582	.3604	.2835	.2230	.1754	.1380	.1086	.0854	.0672	.0528	.0416	15.6
15.8	.4939	.3963	.3181	.2552	.2048	.1644	.1319	.1059	.0850	.0682	.0547	15.8

Note: Blanks indicate P(>t) < 0.00005

Erlang-C Delay Loss Probabilities

(N = 19 Servers)

A	P(>0)	— P(>t) for T_1/T_2 =										A
		0.1	0.2	0.3	0.4	0.5	0.6	0.7	0.8	0.9	1.0	
12.0	.0435	.0216	.0107	.0053	.0026	.0013	.0007	.0003	.0002	.0001		12.0
12.2	.0501	.0254	.0129	.0065	.0033	.0017	.0008	.0004	.0002	.0001	.0001	12.2
12.4	.0575	.0297	.0154	.0079	.0041	.0021	.0011	.0006	.0003	.0002	.0001	12.4
12.6	.0657	.0346	.0183	.0096	.0051	.0027	.0014	.0007	.0004	.0002	.0001	12.6
12.8	.0747	.0402	.0216	.0116	.0063	.0034	.0018	.0010	.0005	.0003	.0002	12.8
13.0	.0847	.0465	.0255	.0140	.0077	.0042	.0023	.0013	.0007	.0004	.0002	13.0
13.2	.0955	.0535	.0299	.0168	.0094	.0053	.0029	.0016	.0009	.0005	.0003	13.2
13.4	.1074	.0614	.0350	.0200	.0114	.0065	.0037	.0021	.0012	.0007	.0004	13.4
13.6	.1204	.0701	.0409	.0238	.0139	.0081	.0047	.0027	.0016	.0009	.0005	13.6
13.8	.1344	.0799	.0475	.0282	.0168	.0100	.0059	.0035	.0021	.0012	.0007	13.8
14.0	.1496	.0907	.0550	.0334	.0202	.0123	.0074	.0045	.0027	.0017	.0010	14.0
14.2	.1659	.1027	.0635	.0393	.0243	.0151	.0093	.0058	.0036	.0022	.0014	14.2
14.4	.1836	.1159	.0732	.0462	.0292	.0184	.0116	.0073	.0046	.0029	.0018	14.4
14.6	.2024	.1304	.0840	.0541	.0348	.0224	.0144	.0093	.0060	.0039	.0025	14.6
14.8	.2227	.1463	.0961	.0632	.0415	.0273	.0179	.0118	.0077	.0051	.0033	14.8
15.0	.2442	.1637	.1097	.0736	.0493	.0331	.0222	.0149	.0100	.0067	.0045	15.0
15.2	.2672	.1827	.1249	.0854	.0584	.0400	.0273	.0187	.0128	.0087	.0060	15.2
15.4	.2916	.2034	.1419	.0990	.0691	.0482	.0336	.0235	.0164	.0114	.0080	15.4
15.6	.3174	.2259	.1608	.1145	.0815	.0580	.0413	.0294	.0209	.0149	.0106	15.6
15.8	.3447	.2503	.1818	.1320	.0958	.0696	.0505	.0367	.0266	.0194	.0141	15.8
16.0	.3736	.2767	.2050	.1519	.1125	.0834	.0618	.0457	.0339	.0251	.0186	16.0
16.2	.4040	.3053	.2307	.1744	.1318	.0996	.0753	.0569	.0430	.0325	.0246	16.2
16.4	.4359	.3361	.2592	.1998	.1541	.1188	.0916	.0706	.0545	.0420	.0324	16.4
16.6	.4694	.3693	.2905	.2285	.1797	.1414	.1112	.0875	.0688	.0541	.0426	16.6
16.8	.5046	.4049	.3250	.2608	.2093	.1680	.1348	.1082	.0868	.0697	.0559	16.8

Note: Blanks indicate P(>t) < 0.00005.

Erlang-C Delay Loss Probabilities

(N = 20 Servers)

A	P(>0)	0.1	0.2	0.3	0.4	0.5	0.6	0.7	0.8	0.9	1.0	A
								P(>t) for T₁/T₂ =				
13.0	.0501	.0249	.0123	.0061	.0030	.0015	.0008	.0004	.0002	.0001	.0001	13.0
13.2	.0572	.0290	.0147	.0074	.0038	.0019	.0010	.0005	.0002	.0001	.0001	13.2
13.4	.0650	.0336	.0174	.0090	.0046	.0024	.0012	.0006	.0003	.0002	.0001	13.4
13.6	.0737	.0389	.0205	.0108	.0057	.0030	.0016	.0008	.0004	.0002	.0001	13.6
13.8	.0832	.0447	.0241	.0129	.0070	.0037	.0020	.0011	.0006	.0003	.0002	13.8
14.0	.0936	.0513	.0282	.0155	.0085	.0047	.0026	.0015	.0008	.0004	.0002	14.0
14.2	.1049	.0587	.0329	.0184	.0103	.0058	.0032	.0018	.0010	.0006	.0003	14.2
14.4	.1172	.0669	.0382	.0218	.0125	.0071	.0041	.0023	.0013	.0008	.0004	14.4
14.6	.1305	.0761	.0443	.0258	.0151	.0088	.0051	.0030	.0017	.0010	.0006	14.6
14.8	.1449	.0862	.0512	.0305	.0181	.0108	.0064	.0038	.0023	.0013	.0008	14.8
15.0	.1604	.0973	.0590	.0358	.0217	.0132	.0080	.0048	.0029	.0018	.0011	15.0
15.2	.1771	.1096	.0678	.0420	.0260	.0161	.0099	.0062	.0038	.0024	.0015	15.2
15.4	.1950	.1231	.0777	.0490	.0310	.0195	.0123	.0078	.0049	.0031	.0020	15.4
15.6	.2140	.1379	.0888	.0572	.0368	.0237	.0153	.0098	.0063	.0041	.0026	15.6
15.8	.2344	.1540	.1012	.0665	.0437	.0287	.0189	.0124	.0081	.0053	.0035	15.8
16.0	.2561	.1717	.1151	.0771	.0517	.0347	.0232	.0156	.0104	.0070	.0047	16.0
16.2	.2791	.1909	.1305	.0893	.0610	.0417	.0285	.0195	.0133	.0091	.0062	16.2
16.4	.3035	.2117	.1477	.1031	.0719	.0502	.0350	.0244	.0170	.0119	.0083	16.4
16.6	.3292	.2343	.1668	.1187	.0845	.0601	.0428	.0305	.0217	.0154	.0110	16.6
16.8	.3564	.2588	.1879	.1365	.0991	.0720	.0523	.0379	.0276	.0200	.0145	16.8
17.0	.3851	.2853	.2113	.1566	.1160	.0859	.0636	.0472	.0349	.0259	.0192	17.0
17.2	.4152	.3138	.2372	.1792	.1355	.1024	.0774	.0585	.0442	.0334	.0252	17.2
17.4	.4468	.3445	.2656	.2048	.1579	.1218	.0939	.0724	.0558	.0430	.0332	17.4
17.6	.4799	.3775	.2970	.2336	.1838	.1445	.1137	.0894	.0704	.0553	.0435	17.6
17.8	.5146	.4129	.3314	.2660	.2134	.1713	.1375	.1103	.0885	.0710	.0570	17.8

Note: Blanks indicate P(>t) < 0.00005

144

Erlang-C Delay Loss Probabilities

(N = 21 Servers)

A	P(>0)	\multicolumn{10}{c}{P(>t) for T_1/T_2 =}	A
		0.1	0.2	0.3	0.4	0.5	0.6	0.7	0.8	0.9	1.0	
14.0	.0567	.0281	.0140	.0069	.0034	.0017	.0008	.0004	.0002	.0001	.0001	14.0
14.2	.0642	.0325	.0165	.0084	.0042	.0021	.0011	.0006	.0003	.0001	.0001	14.2
14.4	.0725	.0375	.0194	.0100	.0052	.0027	.0014	.0007	.0004	.0002	.0001	14.4
14.6	.0816	.0430	.0227	.0120	.0063	.0033	.0018	.0009	.0005	.0003	.0001	14.6
14.8	.0915	.0492	.0265	.0142	.0077	.0041	.0022	.0012	.0006	.0003	.0002	14.8
15.0	.1023	.0562	.0308	.0169	.0093	.0051	.0028	.0015	.0008	.0005	.0003	15.0
15.2	.1140	.0638	.0357	.0200	.0112	.0063	.0035	.0020	.0011	.0006	.0003	15.2
15.4	.1267	.0724	.0413	.0236	.0135	.0077	.0044	.0025	.0014	.0008	.0005	15.4
15.6	.1404	.0818	.0477	.0278	.0162	.0094	.0055	.0032	.0019	.0011	.0006	15.6
15.8	.1551	.0922	.0548	.0326	.0194	.0115	.0068	.0041	.0024	.0014	.0009	15.8
16.0	.1709	.1037	.0629	.0381	.0231	.0140	.0085	.0052	.0031	.0019	.0012	16.0
16.2	.1878	.1162	.0719	.0445	.0275	.0170	.0105	.0065	.0040	.0025	.0015	16.2
16.4	.2059	.1300	.0820	.0518	.0327	.0206	.0130	.0082	.0052	.0033	.0021	16.4
16.6	.2251	.1450	.0934	.0601	.0387	.0249	.0161	.0103	.0067	.0043	.0028	16.6
16.8	.2456	.1614	.1060	.0697	.0458	.0301	.0198	.0130	.0085	.0056	.0037	16.8
17.0	.2673	.1792	.1201	.0805	.0540	.0362	.0243	.0163	.0109	.0073	.0049	17.0
17.2	.2904	.1986	.1358	.0929	.0635	.0434	.0297	.0203	.0139	.0095	.0065	17.2
17.4	.3147	.2196	.1532	.1069	.0746	.0520	.0363	.0253	.0177	.0123	.0086	17.4
17.6	.3404	.2423	.1724	.1227	.0874	.0622	.0443	.0315	.0224	.0160	.0114	17.6
17.8	.3674	.2668	.1937	.1407	.1022	.0742	.0539	.0391	.0284	.0206	.0150	17.8
18.0	.3959	.2933	.2173	.1609	.1192	.0883	.0654	.0485	.0359	.0266	.0197	18.0
18.2	.4257	.3218	.2432	.1838	.1389	.1050	.0793	.0600	.0453	.0343	.0259	18.2
18.4	.4570	.3524	.2717	.2095	.1615	.1245	.0960	.0740	.0571	.0440	.0339	18.4
18.6	.4897	.3852	.3030	.2384	.1875	.1475	.1160	.0913	.0718	.0565	.0444	18.6
18.8	.5239	.4205	.3374	.2708	.2173	.1744	.1400	.1123	.0901	.0723	.0581	18.8

Note: Blanks indicate P(>t) < 0.00005

Erlang-C Delay Loss Probabilities

(N = 22 Servers)

A	P(>0)				P(>t) for $T_1/T_2 =$							A
		0.1	0.2	0.3	0.4	0.5	0.6	0.7	0.8	0.9	1.0	
15.0	.0633	.0314	.0156	.0078	.0038	.0019	.0009	.0005	.0002	.0001	.0001	15.0
15.2	.0713	.0361	.0183	.0093	.0047	.0024	.0012	.0006	.0003	.0002	.0001	15.2
15.4	.0800	.0413	.0214	.0110	.0057	.0029	.0015	.0008	.0004	.0002	.0001	15.4
15.6	.0894	.0472	.0249	.0131	.0069	.0036	.0019	.0010	.0005	.0003	.0001	15.6
15.8	.0997	.0537	.0289	.0155	.0084	.0045	.0024	.0013	.0007	.0004	.0002	15.8
16.0	.1109	.0609	.0334	.0183	.0101	.0055	.0030	.0017	.0009	.0005	.0003	16.0
16.2	.1230	.0689	.0385	.0216	.0121	.0068	.0038	.0021	.0012	.0007	.0004	16.2
16.4	.1360	.0777	.0444	.0253	.0145	.0083	.0047	.0027	.0015	.0009	.0005	16.4
16.6	.1499	.0874	.0509	.0297	.0173	.0101	.0059	.0034	.0020	.0012	.0007	16.6
16.8	.1649	.0981	.0583	.0347	.0206	.0123	.0073	.0043	.0026	.0015	.0009	16.8
17.0	.1810	.1098	.0666	.0404	.0245	.0149	.0090	.0055	.0033	.0020	.0012	17.0
17.2	.1981	.1226	.0758	.0469	.0290	.0180	.0111	.0069	.0043	.0026	.0016	17.2
17.4	.2163	.1366	.0862	.0544	.0344	.0217	.0137	.0086	.0055	.0034	.0022	17.4
17.6	.2357	.1518	.0978	.0630	.0406	.0261	.0168	.0108	.0070	.0045	.0029	17.6
17.8	.2563	.1684	.1106	.0727	.0478	.0314	.0206	.0135	.0089	.0058	.0038	17.8
18.0	.2781	.1864	.1249	.0838	.0561	.0376	.0252	.0169	.0113	.0076	.0051	18.0
18.2	.3011	.2059	.1408	.0963	.0659	.0450	.0308	.0211	.0144	.0098	.0067	18.2
18.4	.3254	.2270	.1584	.1105	.0771	.0538	.0375	.0262	.0183	.0127	.0089	18.4
18.6	.3510	.2498	.1778	.1266	.0901	.0641	.0456	.0325	.0231	.0165	.0117	18.6
18.8	.3779	.2744	.1992	.1447	.1051	.0763	.0554	.0402	.0292	.0212	.0154	18.8
19.0	.4061	.3008	.2229	.1651	.1223	.0906	.0671	.0497	.0368	.0273	.0202	19.0
19.2	.4357	.3293	.2489	.1881	.1421	.1074	.0812	.0614	.0464	.0351	.0265	19.2
19.4	.4666	.3598	.2774	.2139	.1649	.1272	.0981	.0756	.0583	.0449	.0347	19.4
19.6	.4990	.3925	.3088	.2429	.1911	.1503	.1182	.0930	.0732	.0575	.0453	19.6
19.8	.5327	.4275	.3431	.2753	.2210	.1773	.1423	.1142	.0917	.0736	.0590	19.8

Note: Blanks indicate P(>t) < 0.00005

Erlang-C Delay Loss Probabilities

(N = 23 Servers)

A	P(>0)	\multicolumn{11}{c}{P(>t) for T_1/T_2 =}	A
		0.1	0.2	0.3	0.4	0.5	0.6	0.7	0.8	0.9	1.0	
16.0	.0699	.0347	.0172	.0086	.0043	.0021	.0010	.0005	.0003	.0001	.0001	16.0
16.2	.0783	.0397	.0201	.0102	.0052	.0026	.0013	.0007	.0003	.0002	.0001	16.2
16.4	.0873	.0451	.0233	.0121	.0062	.0032	.0017	.0009	.0004	.0002	.0001	16.4
16.6	.0972	.0512	.0270	.0142	.0075	.0040	.0021	.0011	.0006	.0003	.0002	16.6
16.8	.1078	.0580	.0312	.0168	.0090	.0049	.0026	.0014	.0008	.0004	.0002	16.8
17.0	.1193	.0655	.0359	.0197	.0108	.0059	.0033	.0018	.0010	.0005	.0003	17.0
17.2	.1317	.0737	.0413	.0231	.0129	.0072	.0041	.0023	.0013	.0007	.0004	17.2
17.4	.1450	.0828	.0473	.0270	.0154	.0088	.0050	.0029	.0016	.0009	.0005	17.4
17.6	.1592	.0928	.0541	.0315	.0184	.0107	.0062	.0036	.0021	.0012	.0007	17.6
17.8	.1744	.1037	.0617	.0367	.0218	.0130	.0077	.0046	.0027	.0016	.0010	17.8
18.0	.1907	.1157	.0702	.0425	.0258	.0157	.0095	.0058	.0035	.0021	.0013	18.0
18.2	.2080	.1287	.0796	.0493	.0305	.0189	.0117	.0072	.0045	.0028	.0017	18.2
18.4	.2264	.1429	.0902	.0570	.0360	.0227	.0143	.0090	.0057	.0036	.0023	18.4
18.6	.2459	.1583	.1020	.0657	.0423	.0272	.0175	.0113	.0073	.0047	.0030	18.6
18.8	.2665	.1751	.1151	.0756	.0497	.0326	.0214	.0141	.0093	.0061	.0040	18.8
19.0	.2883	.1933	.1295	.0868	.0582	.0390	.0262	.0175	.0118	.0079	.0053	19.0
19.2	.3113	.2129	.1456	.0996	.0681	.0466	.0318	.0218	.0149	.0102	.0070	19.2
19.4	.3355	.2341	.1633	.1139	.0795	.0555	.0387	.0270	.0188	.0131	.0092	19.4
19.6	.3610	.2569	.1829	.1302	.0927	.0659	.0469	.0334	.0238	.0169	.0120	19.6
19.8	.3877	.2815	.2044	.1485	.1078	.0783	.0568	.0413	.0300	.0218	.0158	19.8
20.0	.4157	.3080	.2282	.1690	.1252	.0928	.0687	.0509	.0377	.0279	.0207	20.0
20.2	.4451	.3364	.2542	.1921	.1452	.1098	.0829	.0627	.0474	.0358	.0271	20.2
20.4	.4757	.3668	.2828	.2181	.1681	.1296	.1000	.0771	.0594	.0458	.0353	20.4
20.6	.5077	.3994	.3142	.2471	.1944	.1529	.1203	.0946	.0744	.0586	.0461	20.6
20.8	.5410	.4342	.3484	.2796	.2244	.1801	.1445	.1160	.0931	.0747	.0599	20.8

Note: Blanks indicate P(>t) < 0.00005

Erlang-C Delay Loss Probabilities

(N = 24 Servers)

P(>t) for $T_1/T_2 =$

A	P(>0)	0.1	0.2	0.3	0.4	0.5	0.6	0.7	0.8	0.9	1.0	A
17.0	.0766	.0380	.0189	.0094	.0047	.0023	.0011	.0006	.0003	.0001	.0001	17.0
17.2	.0852	.0432	.0219	.0111	.0056	.0028	.0014	.0007	.0004	.0002	.0001	17.2
17.4	.0946	.0489	.0253	.0131	.0068	.0035	.0018	.0009	.0005	.0002	.0001	17.4
17.6	.1048	.0553	.0291	.0154	.0081	.0043	.0023	.0012	.0006	.0003	.0002	17.6
17.8	.1158	.0623	.0335	.0180	.0097	.0052	.0028	.0015	.0008	.0004	.0002	17.8
18.0	.1275	.0700	.0384	.0211	.0116	.0063	.0035	.0019	.0010	.0006	.0003	18.0
18.2	.1402	.0785	.0439	.0246	.0138	.0077	.0043	.0024	.0014	.0008	.0004	18.2
18.4	.1537	.0878	.0502	.0287	.0164	.0093	.0053	.0031	.0017	.0010	.0006	18.4
18.6	.1682	.0980	.0571	.0333	.0194	.0113	.0066	.0038	.0022	.0013	.0008	18.6
18.8	.1836	.1092	.0649	.0386	.0229	.0136	.0081	.0048	.0029	.0017	.0010	18.8
19.0	.2001	.1213	.0736	.0446	.0271	.0164	.0100	.0060	.0037	.0022	.0013	19.0
19.2	.2175	.1346	.0833	.0515	.0319	.0197	.0122	.0076	.0047	.0029	.0018	19.2
19.4	.2360	.1490	.0941	.0594	.0375	.0237	.0149	.0094	.0060	.0038	.0024	19.4
19.6	.2556	.1646	.1060	.0683	.0440	.0283	.0182	.0117	.0076	.0049	.0031	19.6
19.8	.2763	.1815	.1193	.0784	.0515	.0338	.0222	.0146	.0096	.0063	.0041	19.8
20.0	.2981	.1998	.1339	.0898	.0602	.0403	.0270	.0181	.0122	.0081	.0055	20.0
20.2	.3210	.2195	.1501	.1027	.0702	.0480	.0328	.0225	.0154	.0105	.0072	20.2
20.4	.3452	.2408	.1680	.1172	.0818	.0571	.0398	.0278	.0194	.0135	.0094	20.4
20.6	.3705	.2637	.1877	.1336	.0951	.0677	.0482	.0343	.0244	.0174	.0124	20.6
20.8	.3971	.2883	.2094	.1520	.1104	.0802	.0582	.0423	.0307	.0223	.0162	20.8
21.0	.4249	.3148	.2332	.1728	.1280	.0948	.0702	.0420	.0385	.0286	.0212	21.0
21.2	.4540	.3431	.2593	.1960	.1481	.1119	.0846	.0639	.0483	.0365	.0276	21.2
21.4	.4843	.3734	.2879	.2220	.1712	.1320	.1018	.0785	.0605	.0467	.0360	21.4
21.6	.5159	.4058	.3192	.2511	.1975	.1554	.1222	.0962	.0756	.0595	.0468	21.6
21.8	.5489	.4405	.3535	.2837	.2277	.1827	.1466	.1177	.0944	.0758	.0608	21.8

Note: Blanks indicate P(>t) < 0.00005

Erlang-C Delay Loss Probabilities

(N = 25 Servers)

A	P(>0)	\multicolumn{10}{c}{P(>t) for T_1/T_2 =}	A
		0.1	0.2	0.3	0.4	0.5	0.6	0.7	0.8	0.9	1.0	
18.0	.0831	.0413	.0205	.0102	.0051	.0025	.0012	.0006	.0003	.0002	.0001	18.0
18.2	.0921	.0467	.0236	.0120	.0061	.0031	.0016	.0008	.0004	.0002	.0001	18.2
18.4	.1018	.0526	.0272	.0141	.0073	.0038	.0019	.0010	.0005	.0003	.0001	18.4
18.6	.1123	.0592	.0312	.0165	.0087	.0046	.0024	.0013	.0007	.0004	.0002	18.6
18.8	.1235	.0664	.0357	.0192	.0103	.0056	.0030	.0016	.0009	.0005	.0003	18.8
19.0	.1356	.0744	.0408	.0224	.0123	.0067	.0037	.0020	.0011	.0006	.0003	19.0
19.2	.1481	.0831	.0465	.0261	.0146	.0082	.0046	.0026	.0014	.0008	.0004	19.2
19.4	.1622	.0927	.0529	.0302	.0173	.0099	.0056	.0032	.0018	.0011	.0006	19.4
19.6	.1769	.1031	.0601	.0350	.0204	.0119	.0069	.0040	.0024	.0014	.0008	19.6
19.8	.1925	.1145	.0680	.0405	.0241	.0143	.0085	.0051	.0030	.0018	.0011	19.8
20.0	.2091	.1268	.0769	.0467	.0283	.0172	.0104	.0063	.0038	.0023	.0014	20.0
20.2	.2267	.1403	.0868	.0537	.0332	.0206	.0127	.0079	.0049	.0030	.0019	20.2
20.4	.2453	.1548	.0977	.0617	.0390	.0246	.0155	.0098	.0062	.0039	.0025	20.4
20.6	.2649	.1706	.1099	.0708	.0456	.0294	.0189	.0122	.0078	.0050	.0033	20.6
20.8	.2856	.1877	.1233	.0810	.0532	.0350	.0230	.0151	.0099	.0065	.0043	20.8
21.0	.3074	.2061	.1381	.0926	.0621	.0416	.0279	.0187	.0125	.0084	.0056	21.0
21.2	.3303	.2259	.1545	.1056	.0722	.0494	.0338	.0231	.0158	.0108	.0074	21.2
21.4	.3544	.2472	.1725	.1203	.0840	.0586	.0409	.0285	.0199	.0139	.0097	21.4
21.6	.3796	.2702	.1923	.1369	.0974	.0693	.0494	.0351	.0250	.0178	.0127	21.6
21.8	.4060	.2948	.2141	.1555	.1129	.0820	.0595	.0432	.0314	.0228	.0165	21.8
22.0	.4336	.3212	.2380	.1763	.1306	.0967	.0717	.0531	.0393	.0291	.0216	22.0
22.2	.4624	.3495	.2641	.1996	.1509	.1140	.0862	.0651	.0492	.0372	.0281	22.2
22.4	.4924	.3797	.2928	.2257	.1741	.1342	.1035	.0798	.0615	.0474	.0366	22.4
22.6	.5237	.4120	.3241	.2549	.2005	.1577	.1241	.0976	.0768	.0604	.0475	22.6
22.8	.5563	.4464	.3583	.2875	.2307	.1852	.1486	.1193	.0957	.0768	.0616	22.8

Note: Blanks indicate P(>t) < 0.00005

Appendix D
Engset Loss Probability Tables

The Engset loss probability tables are used to dimension equipment that has a limited number of traffic sources (see Section 3.6). The Engset distribution is based on the following assumptions:

a. finite sources;
b. blocked calls cleared;
c. constant or exponential holding time;
d. full availability.

In the tables, P is used to represent the loss probability, S is used to represent the number of sources, A is used to represent traffic offered to the group in Erlangs, and N is used to represent the number of servers in the group.

Engset Loss Probabilities

(S = 10 Sources)

A	\multicolumn{7}{c}{P for N =}	A
	1	2	3	4	5	6	7	
0.1	.0833	.0034	.0001					0.1
0.2	.1548	.0125	.0006					0.2
0.3	.2166	.0262	.0019	.0001				0.3
0.4	.2705	.0434	.0042	.0003				0.4
0.5	.3178	.0631	.0077	.0006				0.5
0.6	.3596	.0845	.0125	.0012	.0001			0.6
0.7	.3968	.1072	.0186	.0021	.0002			0.7
0.8	.4300	.1305	.0260	.0034	.0003			0.8
0.9	.4599	.1540	.0347	.0052	.0005			0.9
1.0	.4868	.1775	.0446	.0074	.0008	.0001		1.0
1.1	.5113	.2007	.0554	.0103	.0013	.0001		1.1
1.2	.5336	.2234	.0673	.0137	.0019	.0002		1.2
1.3	.5540	.2456	.0799	.0179	.0027	.0003		1.3
1.4	.5727	.2671	.0932	.0226	.0037	.0004		1.4
1.5	.5899	.2879	.1071	.0281	.0050	.0006		1.5
1.6	.6058	.3079	.1214	.0342	.0066	.0008	0001	1.6
1.7	.6206	.3272	.1360	.0410	.0085	.0012	.0001	1.7
1.8	.6343	.3457	.1508	.0484	.0107	.0016	.0002	1.8
1.9	.6470	.3645	.1657	.0564	.0133	.0021	.0002	1.9
2.0	.6589	.3805	.1807	.0650	.0164	.0028	.0003	2.0
2.1	.6700	.3968	.1952	.0740	.0198	.0034	.0004	2.1
2.2	.6805	.4124	.2104	.0835	.0237	.0045	.0006	2.2
2.3	.6903	.4274	.2251	.0934	.0280	.0056	.0007	2.3
2.4	.6995	.4417	.2396	.1037	.0328	.0070	.0010	2.4
2.5	.7082	.4555	.2539	.1143	.0380	.0086	.0012	2.5
2.6	.7164	.4686	.2679	.1251	.0436	.0104	.0016	2.6
2.7	.7242	.4812	.2816	.1361	.0497	.0125	.0020	2.7
2.8	.7315	.4933	.2951	.1472	.0561	.0149	.0025	2.8
2.9	.7385	.5049	.3082	.1585	.0630	.0175	.0031	2.9
3.0	.7451	.5160	.3210	.1698	.0702	.0205	.0038	3.0
3.1	.7514	.5267	.3335	.1811	.0778	.0238	.0047	3.1
3.2	.7574	.5370	.3457	.1924	.0856	.0274	.0057	3.2
3.3	.7631	.5468	.3576	.2037	.0938	.0314	.0068	3.3
3.4	.7686	.5563	.3692	.2149	.1021	.0357	.0081	3.4
3.5	.7738	.5654	.3804	.2260	.1107	.0403	.0096	3.5
3.6	.7789	.5742	.3914	.2371	.1195	.0452	.0113	3.6
3.7	.7836	.5826	.4020	.2480	.1284	.0505	.0132	3.7
3.8	.7881	.5908	.4123	.2587	.1375	.0561	.0154	3.8
3.9	.7925	.5986	.4224	.2694	.1466	.0620	.0177	3.9
4.0	.7967	.6062	.4322	.2798	.1559	.0681	.0204	4.0

Note: Blanks indicate P < 0.00005

Engset Loss Probabilities

(S = 20 Sources)

A	P for N = 2	4	6	8	10	12	14	A
0.2	.0144							0.2
0.4	.0487	.0005						0.4
0.6	.0930	.0020						0.6
0.8	.1410	.0054	.0001					0.8
1.0	.1892	.0112	.0002					1.0
1.2	.2355	.0199	.0006					1.2
1.4	.2791	.0314	.0012					1.4
1.6	.3195	.0457	.0024	.0001				1.6
1.8	.3566	.0624	.0042	.0001				1.8
2.0	.3908	.0811	.0069	.0002				2.0
2.2	.4220	.1014	.0105	.0005				2.2
2.4	.4506	.1228	.0153	.0008				2.4
2.6	.4768	.1449	.0213	.0013				2.6
2.8	.5009	.1674	.0286	.0021	.0001			2.8
3.0	.5231	.1899	.0372	.0032	.0001			3.0
3.2	.5435	.2123	.0470	.0048	.0002			3.2
3.4	.5623	.2343	.0579	.0068	.0004			3.4
3.6	.5798	.2559	.0699	.0093	.0006			3.6
3.8	.5960	.2768	.0828	.0125	.0009			3.8
4.0	.6110	.2971	.0965	.0164	.0013			4.0
4.2	.6250	.3167	.1108	.0211	.0018	.0001		4.2
4.4	.6381	.3356	.1255	.0265	.0026	.0001		4.4
4.6	.6503	.3537	.1406	.0328	.0036	.0002		4.6
4.8	.6618	.3712	.1560	.0398	.0049	.0003		4.8
5.0	.6715	.3879	.1714	.0475	.0066	.0004		5.0
5.2	.6826	.4039	.1868	.0560	.0086	.0006		5.2
5.4	.6922	.4193	.2022	.0652	.0110	.0008		5.4
5.6	.7011	.4340	.2175	.0749	.0140	.0012		5.6
5.8	.7096	.4481	.2326	.0852	.0174	.0016	.0001	5.8
6.0	.7177	.4617	.2474	.0960	.0214	.0022	.0001	6.0
6.2	.7253	.4746	.2620	.1071	.0259	.0029	.0001	6.2
6.4	.7325	.4870	.2763	.1185	.0310	.0039	.0002	6.4
6.6	.7393	.4989	.2902	.1302	.0367	.0050	.0003	6.6
6.8	.7458	.5103	.3039	.1421	.0430	.0065	.0004	6.8
7.0	.7520	.5213	.3172	.1540	.0498	.0082	.0006	7.0
7.2	.7580	.5318	.3301	.1661	.0570	.0102	.0008	7.2
7.4	.7636	.5419	.3427	.1781	.0648	.0127	.0010	7.4
7.6	.7690	.5516	.3550	.1902	.0730	.0155	.0014	7.6
7.8	.7741	.5609	.3669	.2021	.0816	.0187	.0018	7.8
8.0	.7791	.5699	.3785	.2140	.0905	.0224	.0024	8.0

Note: Blanks indicate P < 0.00005

Engset Loss Probabilities

(S = 30 Sources)

A	P for N = 3	5	7	9	11	13	15	A
0.3	.0028							0.3
0.6	.0173	.0002						0.6
0.9	.0450	.0014						0.9
1.2	.0826	.0046	.0001					1.2
1.5	.1259	.0108	.0004					1.5
1.8	.1713	.0210	.0011					1.8
2.1	.2164	.0352	.0026	.0001				2.1
2.4	.2598	.0533	.0052	.0003				2.4
2.7	.3007	.0747	.0094	.0006				2.7
3.0	.3387	.0986	.0154	.0012	.0001			3.0
3.3	.3738	.1243	.0234	.0023	.0001			3.3
3.6	.4061	.1511	.0337	.0040	.0003			3.6
3.9	.4358	.1783	.0460	.0065	.0005			3.9
4.2	.4631	.2054	.0603	.0101	.0009	.0001		4.2
4.5	.4882	.2321	.0763	.0149	.0016	.0001		4.5
4.8	.5113	.2581	.0937	.0209	.0026	.0002		4.8
5.1	.5325	.2833	.1121	.0284	.0041	.0003		5.1
5.4	.5522	.3074	.1314	.0373	.0062	.0006		5.4
5.7	.5703	.3305	.1511	.0476	.0090	.0010	.0001	5.7
6.0	.5872	.3526	.1711	.0591	.0126	.0016	.0001	6.0
6.3	.6028	.3756	.1910	.0718	.0172	.0024	.0002	6.3
6.6	.6174	.3936	.2108	.0854	.0227	.0036	.0003	6.6
6.9	.6309	.4125	.2304	.0998	.0293	.0052	.0005	6.9
7.2	.6436	.4305	.2495	.1149	.0369	.0073	.0008	7.2
7.5	.6555	.4476	.2683	.1304	.0456	.0100	.0013	7.5
7.8	.6666	.4639	.2864	.1462	.0551	.0134	.0019	7.8
8.1	.6771	.4793	.3041	.1622	.0656	.0175	.0028	8.1
8.4	.6869	.4940	.3212	.1783	.0767	.0224	.0039	8.4
8.7	.6962	.5079	.3377	.1943	.0886	.0281	.0055	8.7
9.0	.7050	.5211	.3536	.2101	.1010	.0347	.0074	9.0
9.3	.7132	.5337	.3689	.2258	.1138	.0420	.0099	9.3
9.6	.7211	.5458	.3837	.2412	.1269	.0501	.0129	9.6
9.9	.7285	.5572	.3979	.2563	.1402	.0589	.0165	9.9
10.2	.7356	.5681	.4116	.2711	.1537	.0683	.0208	10.2
10.5	.7423	.5786	.4248	.2855	.1672	.0783	.0257	10.5
10.8	.7486	.5885	.4375	.2996	.1807	.0888	.0314	10.8
11.1	.7547	.5981	.4497	.3133	.1952	.0997	.0377	11.1
11.4	.7605	.6072	.4615	.3267	.2075	.1109	.0446	11.4
11.7	.7760	.6159	.4728	.3396	.2207	.1223	.0521	11.7
12.0	.7713	.6243	.4837	.3522	.2337	.1339	.0603	12.0

Note: Blanks indicate P < 0.00005

Engset Loss Probabilities

(S = 40 Sources)

A	P for N =							A
	3	6	9	12	15	18	21	
0.4	.0064							0.4
0.8	.0355	.0001						0.8
1.2	.0845	.0009						1.2
1.6	.1432	.0035						1.6
2.0	.2038	.0094	.0001					2.0
2.4	.2620	.0198	.0003					2.4
2.8	.3157	.0353	.0010					2.8
3.2	.3643	.0557	.0023					3.2
3.6	.4078	.0801	.0049	.0001				3.6
4.0	.4466	.1075	.0090	.0002				4.0
4.4	.4813	.1369	.0151	.0005				4.4
4.8	.5124	.1672	.0236	.0011				4.8
5.2	.5402	.1977	.0346	.0021	.0001			5.2
5.6	.5653	.2277	.0479	.0038	.0001			5.6
6.0	.5879	.2569	.0635	.0063	.0002			6.0
6.4	.6085	.2851	.0809	.0099	.0004			6.4
6.8	.6271	.3119	.0998	.0148	.0008			6.8
7.2	.6442	.3375	.1198	.0211	.0014			7.2
7.6	.6598	.3617	.1404	.0290	.0024	.0001		7.6
8.0	.6741	.3846	.1615	.0385	.0039	.0002		8.0
8.4	.6874	.4062	.1827	.0494	.0059	.0003		8.4
8.8	.6996	.4267	.2037	.0617	.0088	.0005		8.8
9.2	.7109	.4459	.2245	.0752	.0125	.0008		9.2
9.6	.7214	.4641	.2448	.0896	.0172	.0014		9.6
10.0	.7312	.4812	.2647	.1049	.0230	.0021	.0001	10.0
10.4	.7403	.4974	.2839	.1207	.0299	.0033	.0001	10.4
10.8	.7489	.5127	.3026	.1369	.0380	.0048	.0003	10.8
11.2	.7569	.5272	.3206	.1533	.0471	.0069	.0004	11.2
11.6	.7644	.5409	.3379	.1698	.0572	.0097	.0007	11.6
12.0	.7715	.5539	.3546	.1863	.0682	.0132	.0011	12.0
12.4	.7782	.5662	.3706	.2027	.0799	.0174	.0016	12.4
12.8	.7845	.5779	.3860	.2189	.0922	.0226	.0024	12.8
13.2	.7904	.5890	.4008	.2348	.1051	.0286	.0035	13.2
13.6	.7961	.5996	.4151	.2503	.1183	.0355	.0050	13.6
14.0	.8014	.6096	.4287	.2656	.1317	.0432	.0070	14.0
14.4	.8065	.6192	.4418	.2804	.1454	.0517	.0095	14.4
14.8	.8113	.6284	.4544	.2949	.1591	.0610	.0126	14.8
15.2	.8159	.6371	.4665	.3090	.1728	.0708	.0163	15.2
15.6	.8203	.6455	.4781	.3227	.1864	.0812	.0208	15.6
16.0	.8245	.6534	.4893	.3360	.2000	.0920	.0260	16.0

Note: Blanks indicate P < 0.00005

Engset Loss Probabilities

(S = 50 Sources)

A	\multicolumn{7}{c}{P for N =}	A
	4	8	12	16	20	24	28	
0.5	.0014							0.5
1.0	.0137							1.0
1.5	.0441	.0001						1.5
2.0	.0898	.0006						2.0
2.5	.1438	.0022						2.5
3.0	.2000	.0060						3.0
3.5	.2546	.0133	.0001					3.5
4.0	.3056	.0249	.0003					4.0
4.5	.3522	.0412	.0008					4.5
5.0	.3944	.0618	.0019					5.0
5.5	.4324	.0860	.0039					5.5
6.0	.4666	.1128	.0073	.0001				6.0
6.5	.4974	.1412	.0124	.0002				6.5
7.0	.5251	.1703	.0197	.0005				7.0
7.5	.5501	.1995	.0293	.0010				7.5
8.0	.5729	.2282	.0413	.0020				8.0
8.5	.5935	.2561	.0555	.0035	.0001			8.5
9.0	.6124	.2830	.0716	.0058	.0001			9.0
9.5	.6297	.3088	.0893	.0092	.0002			9.5
10.0	.6455	.3333	.1083	.0138	.0005			10.0
10.5	.6601	.3566	.1280	.0198	.0008			10.5
11.0	.6736	.3786	.1483	.0274	.0015			11.0
11.5	.6861	.3995	.1687	.0365	.0025			11.5
12.0	.6977	.4193	.1892	.0470	.0039	.0001		12.0
12.5	.7085	.4380	.2094	.0589	.0061	.0002		12.5
13.0	.7185	.4556	.2293	.0720	.0090	.0003		13.0
13.5	.7279	.4723	.2487	.0860	.0128	.0005		13.5
14.0	.7367	.4881	.2677	.1008	.0176	.0009		14.0
14.5	.7450	.5031	.2860	.1162	.0236	.0015		14.5
15.0	.7528	.5173	.3038	.1319	.0307	.0024	.0001	15.0
15.5	.7601	.5307	.3209	.1479	.0389	.0036	.0001	15.5
16.0	.7670	.5435	.3375	.1639	.0481	.0054	.0002	16.0
16.5	.7735	.5557	.3534	.1799	.0583	.0077	.0003	16.5
17.0	.7797	.5672	.3687	.1958	.0693	.0108	.0005	17.0
17.5	.7855	.5782	.3835	.2115	.0810	.0146	.0008	17.5
18.0	.7911	.5887	.3977	.2270	.0932	.0193	.0012	18.0
18.5	7963	.5987	.4113	.2421	.1059	.0249	.0019	18.5
19.0	.8013	.6082	.4244	.2569	.1189	.0314	.0029	19.0
19.5	.8061	.6173	.4370	.2714	.1321	.0387	.0042	19.5
20.0	.8107	.6260	.4491	.2855	.1454	.0469	.0060	20.0

Note: Blanks indicate P < 0.00005

Engset Loss Probabilities

(S = 60 Sources)

A	P for N =							A
	4	8	12	16	20	24	28	
0.6	.0026							0.6
1.2	.0241							1.2
1.8	.0709	.0003						1.8
2.4	.1337	.0018						2.4
3.0	.2011	.0063						3.0
3.6	.2661	.0159	.0001					3.6
4.2	.3256	.0319	.0005					4.2
4.8	.3788	.0545	.0015					4.8
5.4	.4258	.0825	.0038					5.4
6.0	.4671	.1144	.0079	.0001				6.0
6.6	.5036	.1485	.0147	.0003				6.6
7.2	.5358	.1835	.0246	.0008				7.2
7.8	.5645	.2182	.0379	.0018				7.8
8.4	.5898	.2519	.0544	.0036	.0001			8.4
9.0	.6126	.2841	.0737	.0066	.0002			9.0
9.6	.6332	.3148	.0952	.0112	.0004			9.6
10.2	.6517	.3436	.1182	.0176	.0008			10.2
10.8	.6685	.3707	.1422	.0261	.0016			10.8
11.4	.6838	.3961	.1665	.0368	.0028	.0001		11.4
12.0	.6978	.4199	.1909	.0495	.0048	.0002		12.0
12.6	.7107	.4421	.2150	.0641	.0078	.0003		12.6
13.2	.7225	.4629	.2386	.0802	.0121	.0006		13.2
13.8	.7334	.4823	.2615	.0975	.0177	.0011		13.8
14.4	.7435	.5006	.2836	.1157	.0248	.0019	.0001	14.4
15.0	.7528	.5176	.3049	.1344	.0335	.0032	.0001	15.0
15.6	.7616	.5337	.3253	.1534	.0438	.0052	.0002	15.6
16.2	.7697	.5488	.3448	.1725	.0554	.0079	.0004	16.2
16.8	.7773	.5629	.3635	.1915	.0682	.0116	.0007	16.8
17.4	.7844	.5763	.3813	.2103	.0819	.0164	.0012	17.4
18.0	.7911	.5889	.3983	.2287	.0965	.0224	.0020	18.0
18.6	.7974	.6008	.4146	.2467	.1116	.0296	.0032	18.6
19.2	.8033	.6121	.4301	.2642	.1272	.0380	.0049	19.2
19.8	.8089	.6228	.4449	.2812	.1429	.0476	.0072	19.8
20.4	.8142	.6329	.4590	.2977	.1588	.0581	.0104	20.4
21.0	.8192	.6425	.4725	.3137	.1746	.0695	.0144	21.0
21.6	.8240	.6516	.4853	.3291	.1902	.0816	.0195	21.6
22.2	.8285	.6603	.4976	.3440	.2057	.0942	.0255	22.2
22.8	.8327	.6686	.5094	.3584	.2210	.1073	.0352	22.8
23.4	.8368	.6765	.5207	.3722	.2359	.1206	.0405	23.4
24.0	.8407	.6840	.5315	.3856	.2506	.1342	.0494	24.0

Note: Blanks indicate P < 0.00005

Engset Loss Probabilities

(S = 80 Sources)

A	\multicolumn{7}{c}{P for N =}	A
	5	10	15	20	25	30	35	
0.8	.0011							0.8
1.6	.0162							1.6
2.4	.0591	.0001						2.4
3.2	.1232	.0009						3.2
4.0	.1949	.0041						4.0
4.8	.2648	.0122	.0001					4.8
5.6	.3285	.0276	.0003					5.6
6.4	.3849	.0507	.0009					6.4
7.2	.4343	.0807	.0027					7.2
8.0	.4773	.1155	.0064	.0001				8.0
8.8	.5149	.1527	.0130	.0002				8.8
9.6	.5479	.1907	.0234	.0005				9.6
10.4	.5770	.2281	.0379	.0014				10.4
11.2	.6027	.2642	.0562	.0030				11.2
12.0	.6256	.2984	.0778	.0060	.0001			12.0
12.8	.6462	.3305	.1019	.0109	.0003			12.8
13.6	.6646	.3606	.1275	.0181	.0006			13.6
14.4	.6813	.3886	.1540	.0278	.0013			14.4
15.2	.6964	.4146	.1806	.0402	.0027			15.2
16.0	.7102	.4388	.2070	.0551	.0049	.0001		16.0
16.8	.7229	.4613	.2328	.0720	.0083	.0003		16.8
17.6	.7345	.4823	.2578	.0905	.0134	.0005		17.6
18.4	.7451	.5018	.2819	.1102	.0202	.0011		18.4
19.2	.7550	.5200	.3049	.1306	.0290	.0020		19.2
20.0	.7641	.5370	.3269	.1513	.0396	.0036	.0001	20.0
20.8	.7726	.5529	.3479	.1722	.0520	.0060	.0002	20.8
21.6	.7805	.5678	.3674	.1929	.0659	.0095	.0004	21.6
22.4	.7879	.5818	.3868	.2132	.0810	.0144	.0008	22.4
23.2	.7948	.5950	.4048	.2331	.0970	.0206	.0014	23.2
24.0	.8013	.6073	.4219	.2525	.1137	.0284	.0025	24.0
24.8	.8074	.6190	.4382	.2713	.1307	.0377	.0041	24.8
25.6	.8131	.6300	.4537	.2894	.1480	.0483	.0065	25.6
26.4	.8185	.6405	.4684	.3069	.1652	.0601	.0099	26.4
27.2	.8236	.6503	.4824	.3238	.1824	.0729	.0144	27.2
28.0	.8284	.6597	.4957	.3400	.1993	.0864	.0201	28.0
28.8	.8329	.6685	.5084	.3556	.2160	.1005	.0270	28.8
29.6	.8373	.6770	.5205	.3706	.2323	.1150	.0351	29.6
30.4	.8414	.6850	.5321	.3851	.2482	.1298	.0444	30.4
31.2	.8453	.6926	.5431	.3989	.2637	.1446	.0546	31.2
32.0	.8490	.6999	.5537	.4122	.2788	.1594	.0657	32.0

Note: Blanks indicate P < 0.00005

Engset Loss Probabilities

(S = 100 Sources)

A	\multicolumn{7}{c}{P for N =}	A
	5	10	15	20	25	30	35	
1.0	.0028							1.0
2.0	.0347							2.0
3.0	.1068	.0006						3.0
4.0	.1958	.0043						4.0
5.0	.2821	.0159	.0001					5.0
6.0	.3582	.0392	.0006					6.0
7.0	.4230	.0740	.0023					7.0
8.0	.4777	.1168	.0069	.0001				8.0
9.0	.5239	.1643	.0162	.0003				9.0
10.0	.5631	.2106	.0315	.0010				10.0
11.0	.5967	.2562	.0529	.0029				11.0
12.0	.6258	.2991	.0795	.0068	.0001			12.0
13.0	.6511	.3388	.1099	.0136	.0004			13.0
14.0	.6733	.3753	.1423	.0242	.0012			14.0
15.0	.6929	.4087	.1754	.0388	.0027	.0001		15.0
16.0	.7103	.4392	.2083	.0571	.0057	.0002		16.0
17.0	.7259	.4670	.2402	.0786	.0107	.0004		17.0
18.0	.7400	.4925	.2709	.1022	.0183	.0011		18.0
19.0	.7527	.5158	.3001	.1273	.0287	.0023	.0001	19.0
20.0	.7642	.5372	.3276	.1530	.0420	.0045	.0002	20.0
21.0	.7747	.5569	.3536	.1789	.0578	.0082	.0004	21.0
22.0	.7843	.5751	.3780	.2044	.0758	.0136	.0009	22.0
23.0	.7932	.5919	.4008	.2294	.0953	.0213	.0018	23.0
24.0	.8013	.6075	.4223	.2535	.1159	.0311	.0034	24.0
25.0	.8089	.6219	.4425	.2768	.1371	.0431	.0060	25.0
26.0	.8158	.6354	.4614	.2991	.1585	.0571	.0100	26.0
27.0	.8223	.6480	.4792	.3204	.1798	.0726	.0156	27.0
28.0	.8284	.6597	.4959	.3407	.2008	.0892	.0230	28.0
29.0	.8341	.6708	.5117	.3600	.2214	.1068	.0322	29.0
30.0	.8394	.6811	.5266	.3784	.2415	.1248	.0431	30.0
31.0	.8444	.6908	.5406	.3959	.2609	.1431	.0555	31.0
32.0	.8490	.7000	.5539	.4126	.2797	.1615	.0692	32.0
33.0	.8534	.7086	.5664	.4285	.2978	.1797	.0838	33.0
34.0	.8576	.7168	.5783	.4436	.3152	.1977	.0992	34.0
35.0	.8615	.7245	.5896	.4580	.3320	.2153	.1149	35.0
36.0	.8652	.7318	.6003	.4717	.3481	.2325	.1310	36.0
37.0	.8688	.7387	.6104	.4848	.3635	.2493	.1471	37.0
38.0	.8721	.7453	.6201	.4973	.3783	.2655	.1632	38.0
39.0	.8753	.7516	6293	.5093	.3926	.2813	.1792	39.0
40.0	.8783	.7576	.6381	.5207	.4063	.2966	.1949	40.0

Note: Blanks indicate P < 0.00005

Glossary

Symbol, Abbreviation, or Acronym	Definition
ATB	all trunks busy
AUTOVON	Military Automatic Voice Network
b/s	bits per second
CAF	call-attempt factor
CCIS	common-channel interoffice signaling
CCITT	International Telephone and Telegraph Consultative Committee
CCS	centum call seconds, hundred call seconds
CCS/MS	CCS per main station
CO	central office
CPU	central processor unit
DDD	direct-distance dialing
DP	dial pulse
DTMF	dual-tone multifrequency
Erl.	Erlang
FIFO	first in, first out
FNPA	foreign numbering plan area
HNPA	home numbering plan area
Hz	Hertz (cycles per second)
IEEE	Institute of Electrical and Electronics Engineers
inst/s	instructions per second
I-O	input-output
ISDN	integrated services digital network
LAN	local-area network
LATA	local access and transport area
MF	multifrequency

MFC	multifrequency compelled
ms	milliseconds
MS	main station
NPA	numbering plan area
ns	nanoseconds
PABX	private automatic branch exchange
PCM	pulse-code modulation
PLF	processor-load factor
PPS	pulses per second
PSTN	public switched telephone network
$P(>t)$	probability of delay greater than indicated time
$P(>0)$	probability call will be delayed
RD	ringdown
SF	single frequency
SLI	serial-line interface
STS	space-time-space
sub.	subscriber
S × S	step-by-step
TDM	time-division multiplex
TO	traffic order
TSI	time-slot interchanger
TST	time-space-time
UC	unit call
VDU	visual display unit
WATS	wide area telecommunication service

Symbol, Abbreviation or Acronym	Definition

Index